Patrick Watson-Williams

Diseases of the Upper Respiratory Tract

The Nose, Pharynx and Larynx

Patrick Watson-Williams

Diseases of the Upper Respiratory Tract
The Nose, Pharynx and Larynx

ISBN/EAN: 9783337179137

Printed in Europe, USA, Canada, Australia, Japan

Cover: Foto ©berggeist007 / pixelio.de

More available books at **www.hansebooks.com**

DISEASES OF THE
UPPER RESPIRATORY TRACT

CORRIGENDA ET ADDENDA.

Page 58, line 32, for *page* 125 read *page* 131.

Page 95, line 24, *cited from Escherich, Wien. Klin. Woch., Jahrg vii., page* 397. For *Fränkel* read *E. Fränkel.*

ILLUSTRATED.

New York:
E. B. TREAT, 5, COOPER UNION.
CHICAGO: 199, CLARKE STREET.

1894.
Reprinted 1895.

DISEASES OF THE
UPPER RESPIRATORY TRACT

THE NOSE, PHARYNX & LARYNX.

BY

P. WATSON-WILLIAMS, M.D., LOND.

*Physician to Out-Patients, and in charge of the Throat Department at the Bristol
Royal Infirmary; Honorary Physician to the Bristol Institute
for the Deaf and Dumb.*

ILLUSTRATED.

New York:
E. B. TREAT, 5, COOPER UNION.
CHICAGO: 199, CLARKE STREET.

1894.
Reprinted 1895.

TO

EDWARD LONG FOX, M.D. Oxon., F.R.C.P.,

AND

MY COLLEAGUES ON THE STAFF OF THE

BRISTOL ROYAL INFIRMARY.

PREFACE.

This Manual is intended to supply a want I have long felt in my capacity as a teacher of clinical medicine, viz., an introduction to the study of disease of the upper air passages that might serve as a practical guide to students and practitioners, and as a stepping stone to the larger and more comprehensive works, whose very wealth of detail renders them less suitable for beginners.

Written by a physician, it deals especially with the medical aspect of rhinology and laryngology, while in order to render it as complete as possible within the prescribed limits, I have endeavoured to place before the reader the opinions of the recognised authorities on the surgical treatment.

Only to a very limited degree can I lay claim to originality, as my purpose has been simply to reflect the present state of knowledge in this department of medicine.

Illustration is essential in describing affections, the diagnosis of which largely depends on the appearances presented, and, with few exceptions, I have reproduced sketches from my clinical notes, believing that they will convey the most accurate idea of the aspects of disease, without elaboration.

The greater portion of this work was written a year ago, but from unavoidable causes its publication has been delayed. This, however, has enabled me to revise the text throughout, and to incorporate advances of knowledge resulting from the International Congress at Rome, and the annual meeting of the British Medical Association at Bristol, this year.

I am under a sense of deep obligation to Dr. McBride, and Mr. Young J. Pentland for permission to reproduce *Fig. 3, Plate IV.*; to Professors Chiari, Emil Zuckerkandl, Dr. Riehl, of Vienna, and to M. Wagnier, of Lille, I am similarly indebted; while I beg also to thank Dr. W. P. Northrup, of New York, Mr. Mark Hovell, the "Scientific Press," and others, for the loan of *clichés*. My especial thanks are due to my friends Dr. James Swain and Dr. Kelly for revising the proof sheets, and to the Publishers for their assistance in preparing the work for the press.

P. W. W.

2, LANSDOWN PLACE,
    CLIFTON, BRISTOL,
October, 1894.

CONTENTS.

CHAPTER I.
INTRODUCTION.
ANATOMY AND PHYSIOLOGY; GENERAL SEMEIOLOGY - - 1—9

CHAPTER II.
EXAMINATION OF THE PHARYNX AND LARYNX.
PHARYNGOSCOPY; LARYNGOSCOPY - - - - 10—20

CHAPTER III.
ACUTE PHARYNGITIS.
ACUTE CATARRHAL PHARYNGITIS; ACUTE PHLEGMON AND ERYSIPELAS OF THE PHARYNX AND LARYNX; RETRO-PHARYNGEAL ABSCESS - - - - - - - 21—29

CHAPTER IV.
CHRONIC PHARYNGITIS.
CHRONIC CATARRHAL, HYPERTROPHIC, AND ATROPHIC PHARYNGITIS; ELONGATED UVULA; VARIX AND ADENOID HYPERTROPHY OF THE TONGUE - - - - - 30—43

CHAPTER V.
DISEASES OF THE TONSILS.
ACUTE TONSILLITIS; CHRONIC HYPERTROPHY OF THE TONSILS; ATROPHIC TONSILLITIS - - - - - 44—53

CHAPTER VI.
CHRONIC INFECTIVE DISEASES.
SYPHILIS, TUBERCULOSIS; LUPUS OF THE PHARYNX AND LARYNX, MYCOSIS - - - - - - 54—63

CHAPTER VII.

NEOPLASMS OF THE PHARYNX AND FAUCES.

BENIGN NEOPLASMS; MALIGNANT NEOPLASMS; DIFFERENTIAL DIAGNOSIS OF SYPHILITIC, TUBERCULAR, LUPOUS, AND MALIGNANT ULCERS - - - - - - 64—70

CHAPTER VIII.

NEUROSES OF THE PHARYNX.

SENSORY NEUROSES; MOTOR NEUROSES - - - 71—72

CHAPTER IX.

ACUTE INFLAMMATIONS OF THE LARYNX.

ACUTE CATARRHAL LARYNGITIS; SPASMODIC LARYNGITIS; ŒDEMA; PERICHONDRITIS OF THE LARYNGEAL CARTILAGES 73—83

CHAPTER X.

CHRONIC LARYNGITIS.

CHRONIC LARYNGITIS, AND PACHYDERMIA LARYNGIS; CHORDITIS TUBEROSA - - - - - - 84—89

CHAPTER XI.

MEMBRANOUS CROUP AND DIPHTHERIA.

MEMBRANOUS CROUP; DIPHTHERIA; INTUBATION - - 90—106

CHAPTER XII.

THROAT AFFECTIONS OF INFECTIOUS FEVERS, GOUT, AND RHEUMATISM.

ENTERIC FEVER, SCARLATINA, MEASLES, SMALL-POX, CHICKEN-POX, INFLUENZA, GOUT, RHEUMATISM - - - 107—120

CHAPTER XIII.

CHRONIC INFECTIVE DISEASES.

SYPHILIS; TUBERCULOSIS; LEPROSY - - - 121—140

CHAPTER XIV.
NEOPLASMS OF THE LARYNX.
BENIGN NEOPLASMS ; MALIGNANT NEOPLASMS - - 141—155

CHAPTER XV.
NEUROSES OF THE LARYNX.
SENSORY NEUROSES ; MOTOR NEUROSES - - - 156—180

CHAPTER XVI.
FOREIGN BODIES IN THE LARYNX. 181—182

CHAPTER XVII.
RHINOSCOPY.
ANTERIOR RHINOSCOPY ; POSTERIOR RHINOSCOPY - - 183—189

CHAPTER XVIII.
RHINITIS.
ACUTE RHINITIS, FIBRINOUS RHINITIS, CHRONIC RHINITIS, HYPERTROPHIC RHINITIS, ATROPHIC RHINITIS, RHINITIS ŒDEMATOSA AND CASEOSA - - - - 190—204

CHAPTER XIX.
HYPERTROPHY OF THE PHARYNGEAL TONSIL.
AND NASO-PHARYNGITIS - - - - - 205—211

CHAPTER XX.
DISEASES OF THE SEPTUM.
PERFORATIONS, DEFLECTIONS AND SPURS, ETC., EPISTAXIS, ABSCESS - - - - - - - 212—220

CHAPTER XXI.
DISEASES OF THE ACCESSORY SINUSES.
THE ANTRUM OF HIGHMORE, THE ETHMOIDAL CELLS, THE FRONTAL SINUS, THE SPHENOIDAL SINUS - - - 221—231

CHAPTER XXII.
CHRONIC INFECTIVE DISEASES.
SYPHILIS; TUBERCULOSIS; LUPUS; GLANDERS; RHINO-SCLEROMA - - - - - - 232—240

CHAPTER XXIII.
NEOPLASMS OF THE NOSE AND NASO-PHARYNX.
MUCOUS POLYPUS, AND BENIGN NEOPLASMS; MALIGNANT NEOPLASMS - - - - - 241—255

CHAPTER XXIV.
NASAL NEUROSES.
OLFACTORY NEUROSES, PARÆSTHESIÆ, NASAL COUGH, VASOMOTOR RHINITIS, HAY FEVER - - - - 256—265

CHAPTER XXV.
FOREIGN BODIES IN THE NOSE. 266—268

FORMULÆ - . - . . . - 269—275

INDEX - . . . . . - 277—282

LIST OF ILLUSTRATIONS.

PLATES.

FRONTISPIECE.—Dissection of the Nose, Pharynx and Larynx.

PLATE I.
- Fig. 1.—The Laryngoscopic Image
- ,, 2.— ,, ,, in Phonation
- ,, 3.— ,, ,, on Deep Inspiration
- ,, 4.—Diagram of the Laryngoscopic Image

Facing p. 16

PLATE II.
- Fig. 1.—Lupus of the Palate and Fauces
- ,, 2.— ,, ,, Larynx and Tongue
- ,, 3.— ,, ,, ,, ,,
- ,, 4.— ,, ,, ,, in early stage

Facing p. 60

PLATE III.
- Fig. 1.—Acute Inflammatory Œdema
- ,, 2.—Laryngeal Tuberculosis, early
- ,, 3.— ,, ,, advanced
- ,, 4.— ,, ,, indurative

Facing p. 80

PLATE IV.
- Fig. 1.—Papilloma of the Larynx
- ,, 2.— ,, ,,
- ,, 3.—Epithelioma of the Larynx
- ,, 4.— ,, ,,

Facing p. 142

PLATE V.
- Fig. 1.—The Anterior Nares
- ,, 2.—The Posterior Rhinoscopic Image
- ,, 3.—Post Nasal Adenoids, etc.

Facing p. 188

WOODCUTS.

FIG.		PAGE
1.—Section of Skull showing Outer Wall of Nasal Fossa		2
2.—The Middle Meatus of the Nose		3
3.—Section of Erectile Tissue of the Inferior Turbinated Body		4
4.—Portable Accumulator for Electric Light and Cautery		11
5.—Türck's Tongue Depressor		12
6.—Diagram to show the proper position of the Mirror in Laryngoscopy		15
7.—Diagram to show the faulty position of the Mirror		18
8.—Fœtal Web of the Glottis		20

LIST OF ILLUSTRATIONS.

FIG.		PAGE
9.—Chronic Pharyngitis		33
10.—Galvano-Cautery Handle and Burners		37
11.—Retraction of Normal Uvula on striking a high note		41
12.—Elongated Uvula		42
13.— ,, ,,		42
14.—Uvula Scissors		42
15.—MacDonald's Guarded Galvano-Cautery		43
16.—Adenoid Hypertrophy at Base of Tongue		43
17.—Tonsillotome, Mackenzie's		49
18.— ,, Reiner's		50
19.—Syphilitic Adhesions of Soft Palate		56
20.—Tubercular Perichondritis of Larynx		79
21.—Subglottic Œdema		79
22.—Syphilitic Perichondritis		82
23.—Perichondritis of the Cricoid Cartilage		82
24.—Laryngorrhœa		85
25.—Pachydermia Laryngis		85
26.—Laryngeal Brush		87
27.—Chorditis Tuberosa		89
28.—Tracheal Probang		100
29.—Intubation Instruments		101
30.—Gouty Pharyngitis		117
31.—Secondary Syphilis of the Larynx		122
32.—Gumma of the Larynx		123
33.—Syphilitic Larynx with Outgrowths		123
34.—Cicatricial Contraction of Syphilitic Larynx		124
35.—Calomel Fumigator		125
36.—Störk's Tracheal Cannula		126
37.—Laryngeal Tuberculosis		128
38.— ,, ,,		128
39.— ,, ,, Advanced		128
40.—Tubercular Ulceration of Vocal Cords		130
41.—Tubercular Larynx		130
42.—Author's Syringe for Submucous Injections in Laryngeal Tuberculosis		133
43.—Krause's Cutting Curettes		134
44.— ,, Forceps for applying Lactic Acid to the Larynx		134
45.—Heryng's Laryngeal Curettes		135

LIST OF ILLUSTRATIONS.

Fig.		Page
46.—Leprosy of the Larynx		138
47.— ,, ,,		139
48.—Papilloma of the Larynx		142
49.—Cystoma of Larynx		143
50.—Mackenzie's Ecraseur for Laryngeal growths		145
51.—Laryngeal Forceps, Mackenzie's Lateral		146
52.— ,, ,, ,, Antero-posterior		146
53.— ,, ,, Schrötter's		146
54.— ,, Snare, Gibb's		147
55.—Epithelioma of the Epiglottis		149
56.—Laryngeal Insufflator		155
57.—Unilateral Paralysis of Superior Laryngeal Nerve		165
58.—Bilateral ,, ,, ,, ,,		165
59.—Paralysis of Thyro-Arytenoidei Muscles		165
60.— ,, ,, Interni Muscles		166
61.—Arytenoideus, Diagram of		167
62.— ,, Paralysis of		167
63.—Paralysis of Arytenoideus and Thyro-Arytenoids		167
64.—Abductors of Vocal Cords ; Diagram of		168
65.— ,, ,, ,, Bilateral Paralysis of		169
66.—Adductors of Vocal Cords ; Diagram of		171
67.— ,, ,, ,, Bilateral Paralysis of		171
68.—Paralysis of Vocal Cord on Phonation		172
69.— ,, ,, ,, Deep Inspiration		172
70.—Ankylosis of Crico-Arytenoid Joint on Phonation		174
71.— ,, ,, ,, Deep Inspiration		174
72.—Coin in Larynx		181
73.—Wolfenden's Forceps		181
74.—Duplay's Nasal Speculum		183
75.—Lennox Browne's Nasal Speculum		183
76.—Fränkel's Nasal Speculum		184
77.—Neil Griffith's Nasal Speculum		184
78.—Zaufal's Speculum		185
79.—Michel's Rhinoscope		186
80.—Palate Retractor, Voltolini's		187
81.— ,, ,, Cresswell Baber's		187
82.—Diagram of Rhinoscopic Image		188
83.—Wilkin's Polypus Snare		196

Fig.		Page
84.	—Galvano-caustic Snare for Nose	196
85.	—Laryngo-tracheal Ozæna	200
86.	—Nasal Douche	202
87.	—Normal Membrana Tympani	205
88.	—Retraction of Membrana Tympani	206
89.	—Eustachian Synechiæ	208
90.	—Lennox Browne's Finger Guard and Curette	209
91.	—Dalby's Scraper for Post-Nasal Adenoids	209
92.	—Mark Hovell's Forceps for Post-Nasal Adenoids	209
93.	—Gottstein's Curette ,, ,, ,,	210
94.	—Hartmann's ,, ,, ,, ,,	210
95.	—Bronner's ,, ,, ,, ,,	210
96.	—Septal Deflection and Turbinal Hypertrophy	213
97.	— ,, ,, ,, ,, after Cocaine	213
98.	— ,, ,, etc.	214
99.	—Bosworth's Nasal Saw	215
100.	—MacDonald's Gouges for Septal Spurs	215
101.	—Curtis' Trephines ,, ,,	216
102.	—Major's Nasal Knife ,, ,,	216
103.	—Hill's Dilator	217
104.	—Jarvis' Steel Septum Forceps	217
105.	—Transverse Section of Nasal Cavities and Maxillary Sinuses	223
106.	—Lichtwitz Trocar	225
107.	—Section showing Irregular Development of Maxillary Sinuses	226
108.	—Superior Maxilla, Trephined for Antral Empyema	227
109.	—Section of Mucous Polypus of Nose	242
110.	—Moriform Hypertrophy of the Inferior Turbinated Bodies	245
111.	—Nasal Polypus Punch Forceps	247
112.	— ,, ,, Alligator Forceps	248
113.	— ,, ,, Forceps	248
114.	—Polypus Snare, Hovell's	249
115.	— ,, ,, Krause's	250
116.	—Fibroma of the Naso-Pharynx	252
117.	—Papilloma of the Septum	252
118.	—Diagram of Nerve Supply to the Nasal Passages	258
119.	—Oil Atomiser	273
120.	—Spray	274

DISEASES OF THE UPPER RESPIRATORY TRACT.

CHAPTER I.

INTRODUCTION.

ANATOMY AND PHYSIOLOGY; GENERAL SEMEIOLOGY.

ANATOMY AND PHYSIOLOGY.

IN a work addressed to medical practitioners and senior students it may be assumed that every reader is acquainted with the topography of the regions under discussion, and has at hand the usual text-books of anatomy and physiology; yet it is desirable to draw attention to a few points of special importance from a *clinical* aspect.

The anatomical relations of the upper respiratory tract are well shown in *the frontispiece*, from a dissection in the museum of the Bristol Medical School, which may be studied in connection with the description of the several structures as they appear on examination by *rhinoscopy* and *laryngoscopy*.

To show the structure of the nasal fossæ I have cut three sides of the *septum nasi*, and raised it as a lid, and have passed bristles through the openings in the walls of the fossæ, by means of which the accessory sinuses communicate with the nose. This plate, being an exact reproduction from nature, will afford precise information as to the relations of the different regions of the nose, pharynx, and larynx, and will be found very useful to the beginner as an aid to clearly comprehending the directions for performing various manipulations necessary for the diagnosis and treatment of disease of the nose and throat.

In order to give a better idea of the position and size of the "hiatus semilunaris," or orifice by which the infundibulum, the anterior ethmoidal cells and the antrum of Highmore communicate with the middle

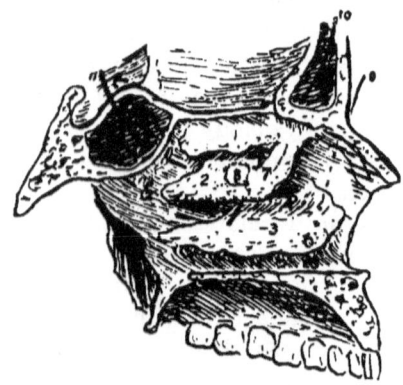

FIG. 1.

Section of skull, showing the outer wall of the left nasal fossa: 1, the superior, 2, middle, and 3, inferior turbinated bones; 7, dotted outline showing the position of the hiatus semilunaris, and 8, of the ostium maxillare beneath the middle turbinated bone; 9, a bristle in the nasal duct; 10, a bristle passed into the infundibulum; 11, a bristle, passed through the opening of the sphenoidal cells.

meatus of the nose, I have drawn up the middle turbinated body (2) by means of a hook, and have had these portions of the dissection reproduced in their natural size in the illustration on the next page. This region is one of great importance in nasal disease.

The physiological functions of the nose are threefold, vocal, olfactory, and respiratory; but I need only refer here to the latter to emphasise its important bearing on many diseases of the respiratory tract, of which the nasal passages constitute the first portion.

We know that "mouth-breathing" is not only a source of inconvenience, but is generally attended with troublesome affections of the lower respiratory tract. The nose therefore performs important functions which cannot be fulfilled by the mouth, and a careful examination of its structure affords a ready explanation of these functions.

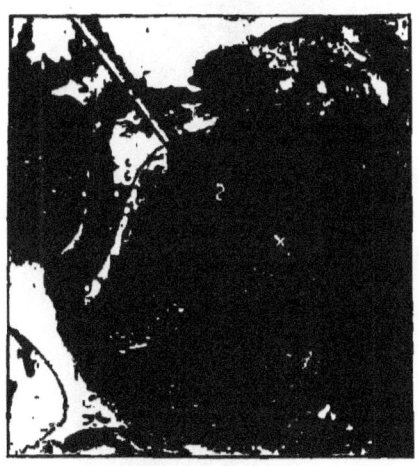

FIG. 2.
A part of the dissection (frontispiece) enlarged to the natural size. The middle turbinated body (1) has been raised by a hook, exposing (2) the middle meatus and showing the hiatus semilunaris immediately behind the bulla ethmoidalis x.

The superior half of the nasal fossæ is mainly concerned with the function of smelling: but the mucous membrane of the middle and inferior turbinated bodies contains vascular erectile tissue, which warms and moistens the inspired air, and in these regions consists of connective tissue, the surface of which is covered with ciliated epithelium, the deep portion forming the periosteum. Between these two layers are

4 *THE UPPER RESPIRATORY TRACT.*

lymphoid tissue, an abundant supply of lymphatics, and numerous venous plexuses, into which the capillary veins freely open, as shown in the accompanying illustration of a section of a hypertrophied inferior turbinal removed by snaring.

Around the venous plexuses unstriped muscular tissue is distributed. When the plexuses are filled the turbinal bodies swell considerably, but numerous elastic

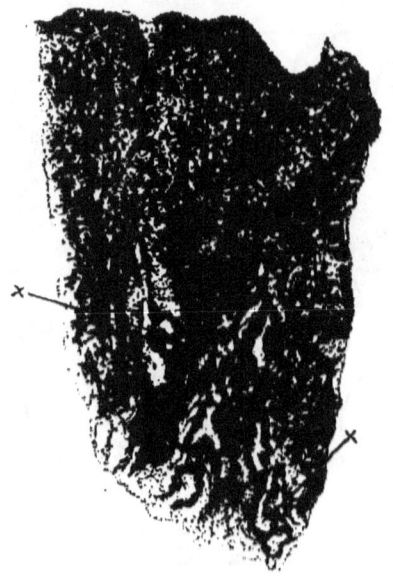

FIG. 3.
Section of hypertrophied erectile tissue from the inferior turbinated body; showing the arrangement of the vessels and venous spaces into which the collecting veins are seen to open.

fibres in the deeper layers cause the tissue to collapse, unless actively distended by the peculiarly arranged vascular supply. Similar erectile tissue is found occupying the lower part of the septum opposite the anterior extremity of the middle turbinal and the floor of the nasal fossæ.

It can readily be conceived that these highly vascular mucous tissues are capable of rapidly warming inspired air as it passes over them, and of secreting with equal rapidity a copious supply of watery mucus which can yield its moisture to the warmed air. That this actually does take place has been demonstrated by Gréhant, Kayser, Bloch, MacDonald and others.

The conclusions arrived at by these observers may be summarised as follows :—

1.—Whatever the temperature of the atmosphere, the air, after ordinary inspiration through the normal nasal passages, is always raised to the temperature of the blood before reaching the pharynx.

2.—The air, after passing through the nose, is invariably completely saturated with moisture.

It has been estimated that about half a pint of water is taken up by the inspired air in the course of twenty-four hours, and, if the air is to reach the lungs in the normal condition of saturation, it is obviously impossible for that amount of moisture to be taken up from the bronchial mucous membrane without serious risk to its physiological integrity. We know that we can breathe normally all the night through, and awake in the morning with the pharynx and larynx in a moist and healthy condition ; but if we have lain an hour or two breathing through the open mouth, the pharynx is quite dry, or is covered with a tenacious thick mucus, and if this mouth-breathing becomes a habit, the pharynx and larynx suffer from chronic congestion, and the parts are "relaxed." It is, in fact, unnecessary to dilate on the pernicious effects of mouth-breathing ; the constant tendency to laryngitis, attacks of bronchitis, and various other ways in which the evil effects are manifested, are familiar to practitioners.

GENERAL SEMEIOLOGY.

Before passing on to the detailed description of diseases of the pharynx, larynx and nose, it may be helpful to review, very briefly, some of the more general symptoms associated with disease in these regions, in order to direct attention to the very important part that general therapeutical measures should occupy in their treatment, and to emphasise the desirability of examining the pulse, heart, lungs, etc., and of investigating the personal and family history of patients complaining of their throat and nose, so as to avoid the error of having regard only to local manifestations of general affections.

Thus gout, rheumatism, and dyspepsia, are very frequent causes of pharyngitis and laryngitis, and of certain forms of rhinitis, and though from long standing congestion local treatment may be required, the great majority of cases yield only to appropriate general treatment.

As an instance of the importance of taking into consideration the general condition of a patient's health, I may cite the case of a clergyman sent to me by his medical attendant for painful sore throat and loss of voice. He was the Vicar of a large country parish, and was thoroughly overworked by his Lenten services. The fauces were intensely injected, and he complained of severe pain, especially in swallowing. The larynx was greatly congested, and he was almost voiceless.

But on feeling his pulse, I found that it was frequent, 96, small and irregular, while the heart beats were 136 to the minute! There was no history of gout or rheumatism, yet the patchy appearance of the faucial and laryngeal congestion confirmed my suspicions that an indefinite swelling about one elbow joint was gouty, and the rapid improvement under appropriate general treatment that I was led to suggest to his medical attendant, with the help of a sedative spray and pastil, very soon restored him to health. It is scarcely necessary to point out that to have had regard only to the throat condition in a patient whose heart's action was so weak and feeble that nearly every other beat failed to reach the wrist, would have been disastrous.

Again, a slight persistent laryngitis may be the earliest symptoms of tubercular disease of the lungs, and should lead us to examine the chest and feel the pulse.

Paralysis of a vocal cord may be caused by pressure of an aneurism on a recurrent laryngeal nerve, and is often the earliest indication of an intra-thoracic tumour, or it may be due to some central nerve lesion, such as locomotor ataxia, bulbar paralysis, nuclear disease from syphilis, or to an intracranial growth.

Intra-nasal disease is the exciting factor in a large percentage of spasmodic neuroses, *e.g.*, hay-fever, persistent cough, asthma, or even epilepsy, migraine, etc.

Patients generally complain of either pain, impairment of voice, respiratory obstruction, difficulty in swallowing, or increased secretion.

Pain may be very considerable in simple dyspeptic or gouty pharyngitis, in malignant disease, and in tuberculosis, but, on the other hand, it is often almost absent in early malignant disease, and is usually slight in syphilis and lupus.

Vocal impairment may be due to thickening of the pharyngeal mucous membrane, or perforation of the velum palati, or the voice may be nasal in character from nasal obstruction. The cause of vocal alteration may be in the larynx and amount to hoarseness, or loss of voice (*aphonia*), or may arise from ulceration, inflammatory thickening of the vocal cords, imperfect approximation of the cords from growths, thickening in the inter-arytenoid fold, or from paresis of the vocal cords.

It acquires a peculiar raucous character in old syphilitic disease, which is almost pathognomonic.

On the other hand, the voice may be quite unaltered in the most dangerous form of vocal cord paralysis, viz., bilateral paralysis of the abductors, and hardly appreciably altered in paralysis of one vocal cord.

Obstruction to respiration results from nasal obstruction in rhinitis, polypus, growths or deviation of the septum, and when the respiratory difficulty lies in the nasal passages, it is sure to cause various pathological conditions in the naso-pharynx, if not also in the larynx. Nasal respiration must be carefully investigated, for in incomplete obstruction, the respiration may appear to be conducted normally, so long as the patient remains at rest, and thus even considerable degrees of nasal obstruction may escape notice. Deposits of false membrane, laryngeal growths, œdema of the larynx, or perichondrial thickening, may encroach on the glottic space, and give rise to very marked dyspnœa, which may be absent, if only the epiglottis, or the epiglottic folds, are involved. Dyspnœa, of course, results from spasm of the glottis. The only paralysis of the vocal cords which would give rise to great dyspnœa is bilateral paralysis of the abductors, an extremely rare affection.

The obstruction may be infraglottic, and due to infraglottic growths, or œdema, or to a foreign body, or to pressure on the trachea by a neighbouring growth or aneurism.

Deglutition may be difficult, *dysphagia ;* painful, *odynphagia ;* or absolutely impossible, *aphagia.*

Dysphagia may arise from paresis of the constrictors, as in bulbar paralysis, from obstruction from a growth in the pharynx or œsophagus, from pressure of a growth or aneurism on the pharynx or œsophagus, or it may be due to a cicatricial stenosis or to spasm.

Odynphagia is commonly experienced in gouty, rheumatic, and dyspeptic pharyngitis, and in malignant disease, is generally present in tubercular disease of the pharynx, epiglottis or arytenoid regions, and is usually considerable in perichondritis of the cricoid cartilage.

Aphagia results from complete obstruction of the laryngo-pharynx or œsophagus.

Increased secretion from the fauces and larynx is common to all catarrhal affections, malignant disease, and tubercular disease, and sometimes to a less extent in syphilis.

Nasal discharge is increased and is generally muco-purulent in all forms of rhinitis, except the atrophic form, and is also seen in hay-fever, vaso-motor rhinitis, polypus, and in syphilitic, tubercular and lupous disease. A discharge of pus from the nose may come from one of the accessory sinuses.

If fœtid, it is probably due to syphilis, ozæna, malignant disease, or antral empyema. The patient is unconscious of the fœtor in atrophic rhinitis, and may not perceive the smell when syphilitic necrosis is present, but the fœtid odour of antral empyema is perceived by the patient himself.

The well known physiognomy of a patient with syphilitic disease which has involved the bony structures hardly requires description here, but the peculiar facial appearance of a child with post-nasal adenoids is equally characteristic. The *broadened* saddle-back bridge of the nose, giving the appearance of a sunken bridge, is due to infiltration of the subcutaneous tissues in post-nasal adenoids, while the same appearance results in old syphilitic disease from cicatricial contraction of these tissues, even in cases in which there is no actual falling in of the bony structures.

Chapter II

EXAMINATION OF THE PHARYNX AND LARYNX.

PHARYNGOSCOPY; LARYNGOSCOPY.

In addition to the necessary apparatus supplied by the instrument makers, *viz.*, (1,) a concave reflecting forehead mirror; and (2,) the laryngeal mirror—about the size of a half-penny, and connected with a handle—the source of light is a matter for consideration; and for satisfactory examination a brilliant light is essential. While bright sun-light answers admirably when it is available, it is generally more convenient to employ artificial light, which is more under control. Any good lamp will do, but special lamps give a much better illumination of the parts to be examined, and for posterior rhinoscopy are an absolute necessity. Without attempting to describe in detail the many forms of lamps and brackets, I may mention Morell Mackenzie's gas bracket, with an argand burner and a bull's-eye condenser to collect the rays as one of the most cleanly and convenient. If fitted with an eighty-candle power Welsbach incandescent burner, with the adjustment for rotating it on its axis, it should cost under £3, and we then have a light which, for brilliancy and whiteness, is only excelled by the oxyhydrogen lime-light. If cost is no great obstacle, the oxyhydrogen lime-light offers some advantages over the Welsbach.

It is very essential that the light should be freely moveable in every direction, so as to allow of ready adjustment in focussing the light on the part to be examined. If Semon's laryngeal mirror with a small electric lamp attached be used, the forehead mirror and all other lamps may be dispensed with. A small portable accumulator, weighing only a few pounds, will keep such a lamp illuminated for several hours in the aggregate, and is very convenient, especially for use at patients' houses. But when exhausted, it can only be re-charged by an electrician having a suitable dynamo, and in the country some trouble and delay would be involved in getting the re-charging done. The actual cost of re-charging is insignificant.

FIG. 4.

A convenient Portable Accumulator for electric light and cautery.

The reflecting mirror is of about 14-inch focus, although one of shorter focus, similar to that for examining the ear, is sometimes recommended for examining the nose. It should have an opening in the centre, through which the eye of the observer looks. It is fixed to the forehead by means of a band round the head or a metal band over the head, such as Fox's head band, or carried on a spectacle frame. It is essential that the central opening should come immediately in front of the observer's corresponding pupil, and that the mirror should be freely adjustable.

PHARYNGOSCOPY.

In examining the pharynx we sit facing the patient with the reflecting mirror adjusted either in the front of the centre of the forehead or over the right eye, in such a position that the eye looks through the aperture in the centre. The lamp is placed on either side, usually the patient's left, about four inches from the patient's head, on a level with his ear, and so that the light is directed to the reflecting mirror on the examiner's head, and thence into the patient's mouth. It is generally necessary to depress the tongue either with a tongue depressor, spoon, or spatula. Fränkel's is a very handy tongue depressor. The under surface of the extremity is serrated so as to grasp the tongue, while the fenestra may be utilised to draw forward the uvula in inspection of the posterior pharyngeal wall. Türck's will be found useful in some cases, as the handle is twisted to the left of the median line so that the left hand does not come in the way of the right when using the rhinoscope. In using the tongue depressor it should be placed well over the arch or dorsum of the tongue, and at first only gentle pressure should be used. If we attempt to suddenly and forcibly depress the organ, it will be involuntarily arched up.

FIG. 5.
Türck's Tongue Depressor.

For clinical purposes the pharynx is divided into three regions, the naso-pharynx, the oro-pharynx and the laryngo-pharynx.

The *naso-pharynx* is continuous with the anterior nasal cavities, and extends from the base of the occiput and sphenoid downwards as far as the isthmus, the narrow space corresponding to a line drawn from the posterior margin of the soft palate to the posterior pharyngeal wall. Into it open the Eustachian tubes by their trumpet-shaped orifices, from the posterior margin of which may be seen the salpingo-pharyngeal folds extending downwards, and forming on each side a fossa between them and the posterior wall of the pharynx, the fossa of Rosenmüller. The orifices of the Eustachian tubes are just behind the posterior extremity of the inferior turbinated body.

The mucous membrane is covered with ciliated columnar epithelium, and is more abundantly supplied with mucous glands than the anterior nasal cavities. Numerous lymphoid follicles exist throughout the pharynx, and a collection of these in the roof and posterior wall of the naso-pharynx forms a mass, similar to the faucial tonsils, termed Luschka's, or the pharyngeal tonsil. The pharyngeal tonsil presents an uneven surface with longitudinal ridges. At the lower extremity is the elevated *bursa pharyngea*, with a central depression, the "foramen."

The *oro-pharynx* lies between the isthmus above, and a horizontal line on a level with the top of the epiglottis.

The *laryngo-pharynx* extends down to a point corresponding with the cricoid cartilage where the œsophagus commences, and includes the hyoid fossæ and parts generally described with the larynx.

The tonsils are simply collections of lymphoid tissue lying between the anterior and posterior pillars of the fauces.

First, the condition of the parts during quiet breathing should be noted. The tonsils are seen lying between the pillars of the fauces, and though they should not normally project beyond the free margin of the faucial pillars, it is by no means uncommon to find some enlargement in persons who do not complain of any throat trouble whatever. Similarly a considerable variation from the usual pink, smooth character of the mucous membrane of the pharynx must be regarded as within the limits of the normal. The patient should then be directed to sound *ah!* and the retraction of the soft palate noted. While noting the colour and contour of the *velum palati*, we must guard against overlooking any diseased condition on its posterior surface, for, especially in syphilis, extensive ulceration may affect its posterior surface only, and nothing strikingly abnormal appear on the anterior; yet on careful observation we always find sufficient diffuse hyperæmia with deficient mobility of the soft palate to put us on our guard. The soft palate and uvula are rich in small mucous glands, and sometimes these appear as small elevations the size of millet seeds. It is desirable to make the patient "gag" or retch before the examination is concluded, as by this means the pharyngeal muscles bring into view the lateral walls behind the tonsils, disclosing any thickening such as is found in *pharyngitis lateralis* and sometimes showing a considerable hypertrophy of the tonsils previously unobserved.

Among congenital deformities that may be encountered we may mention perforation of the anterior faucial pillar, bifid uvula, absent uvula or soft palate, and divided soft palate (cleft palate).

The examination of the naso-pharynx is described under the examination of the nose, and need not be referred to now.

LARYNGOSCOPY.

The *face* of the laryngeal mirror must be warmed over the lamp, so as to prevent the breath from condensing on it and obscuring the image.

The patient, with the head slightly thrown back, should then open the mouth widely and, breathing quietly, protrude the tongue, which is gently but firmly held in a small towel by the physician's left hand. The laryngeal mirror should be lightly held in the examiner's right hand as one holds a pen, steadily but gently pressing against the uvula and soft palate, yet not so far back as to actually touch the posterior wall of the pharynx.

At first, only the back of the tongue and the epiglottis will be seen reflected in the laryngeal mirror, but by altering its angle the other parts will be successively brought into view. While keeping the mouth widely open, the patient should be directed to sound *eh! eh!* thus causing the larynx to be raised and brought more readily into view. The vocal cords can then be seen approaching and diverging alternately in phonation and inspiration.

It will be noticed that the laryngeal image is inverted, since it is seen reflected in the laryngeal mirror. The reason for this inversion will be obvious from the accompanying diagram, which shows that the anterior or upper

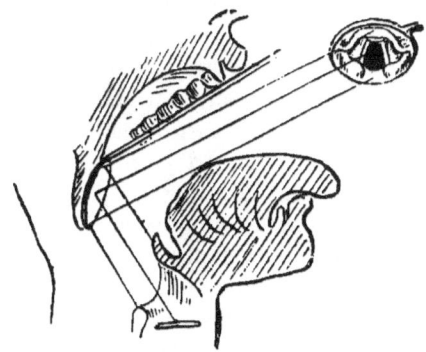

FIG. 6.
Diagram to show the position of the laryngoscopic mirror which will give the most perfect view of the larynx, and to explain the inversion of the laryngeal image.

portion of the mirror is lying immediately over the extreme base or posterior portion of the tongue and epiglottis, while the posterior or lower part of the mirror is above the posterior half of the larynx. Thus it is that the anterior epiglottis is seen reflected in the upper half of the mirror. The relative position of the parts on the right and left remains, of course, unaltered.

Clinically, we distinguish three portions of the larynx,

(1,) the supra-glottic, or the part above the ventricular bands; (2,) the glottic part, between the ventricular bands and the vocal cords, including the ventricles, and (3,) the infra-glottic, extending from the vocal cords to the lower border of the cricoid cartilage.

The laryngoscopic image brings into view the following structures. (See *Plate I.*, *Figs.* 3 and 4).

The most striking object is the epiglottis, which, from its being bent on itself, shows portions of both the upper and lower surfaces.

The epiglottis normally varies greatly in shape and position in different patients. In some it is erect and only slightly curved (*Figs.* 1, 3); in others it overhangs the larynx, or is very much curved and curled. (See *Plate IV.*).

Behind the epiglottis are the diverging pearly white vocal cords passing backwards to be attached to the arytenoid cartilages. The cartilages of Wrisberg and Santorini are seen as rounded swellings in the lower part of the image, and forming in part the posterior boundary of the larynx and between the arytenoid cartilages is the inter-arytenoid space. The ary-epiglottic folds of mucous membrane pass from the arytenoid cartilages, on either side, forward to the epiglottis.

The true vocal cords, essentially the ligamentous portion of the thyro-arytenoid muscles, are attached posteriorly to the *processi vocales* and to the anterior surface of the arytenoid cartilages, and, passing forwards, are attached in front in the angle of the thyroid cartilage, forming the anterior commissure, just below the projection or thickening termed the cushion of the epiglottis.

Between the true vocal cords is the opening of the larynx, the *rima glottidis*, and external to the vocal cords, and lying on a higher level are the pink false

PLATE I.

The Normal Larynx.

Fig. 1.

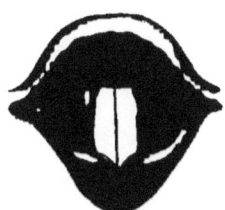

Fig. 2.

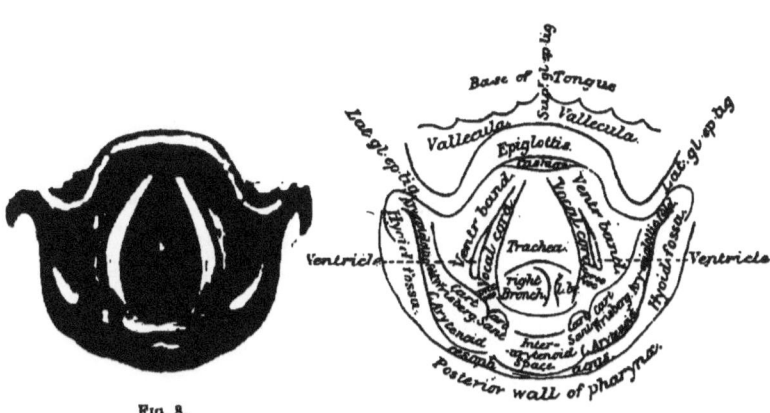

Fig. 3.

Fig. 4.

Fig. 1.—Laryngoscopic image of a normal larynx during quiet respiration. The vocal cords are lying midway between adduction and abduction,—the *so-called* "cadaveric position."

Fig. 2.—The same during vocalisation. The vocal cords are adducted and the arytenoid cartilages are brought into apposition.

Fig. 3.—A normal larynx on deep inspiration showing the vocal cords widely abducted.

Fig. 4.—Explanatory diagram of the laryngoscopic image.

cords or ventricular bands. In some cases, especially by tilting the mirror laterally, the opening of the *sacculus laryngis*, or ventricle of Morgagni, can be seen on either side between the ventricular band and the cord. A few rings of the tracheal cartilages can generally be seen below the vocal cords, and rarely, by directing the light well down through the rima, the division between the right and left bronchi may be made out.

The epiglottis is attached to the base of the tongue by three ligamentous folds, the one central (superior glosso-epiglottic lig.), and the right and left lateral glosso-epiglottic folds. The spaces between these folds are named the *valleculæ*.

The opening of the upper end of the œsophagus lies between the posterior margin of the larynx and the posterior wall of the pharynx, but though shown as an actual opening in the diagram, the larynx lies in contact with the posterior wall of the pharynx, except during the passage of food. On either side of the larynx are the hyoid fossæ.

In making a laryngeal examination, we first observe the larynx during quiet respiration (*Plate I. Fig.* 1), noting whether the colour of the mucous membrane is healthy, and the form of the various structures normal, and free from swelling or ulceration. The epiglottis is slightly yellowish, and the rest of the laryngeal mucous membrane is pale pink. The vocal cords should be pearly white, or very slightly pink, and the free margins perfectly smooth and even. They should lie symmetrically midway between abduction and adduction, commonly called the cadaveric position because it was thought to be the position assumed by the vocal cords in the dead body. Semon has shown, however, that as a matter of fact in the position of rest, owing to reflex-tonus of the abductors, the glottic chink is wider than in the cadaver, or after section of the recurrent laryngeal nerves.

Next we note the movements of the vocal cords during vocalisation (*Plate I, Fig.* 2), and deep inspiration (*Plate I, Fig.* 3). On phonating *eh! eh!* the vocal cords should approach till the free margins almost meet in the middle line, the arytenoid cartilages at the same time being also approximated by the arytenoideus muscle so as to obliterate the inter-arytenoid space.

In deep inspiration the vocal cords are widely abducted, and diverge considerably more than during quiet respiration, the inter-arytenoid space being correspondingly increased.

Difficulties in Laryngoscopy.—We may fail to obtain a good laryngoscopic image, either from faulty manipulation, or owing to the peculiar conformation of the laryngeal structures :—

(1,) A common fault is to hold the laryngoscopic mir-

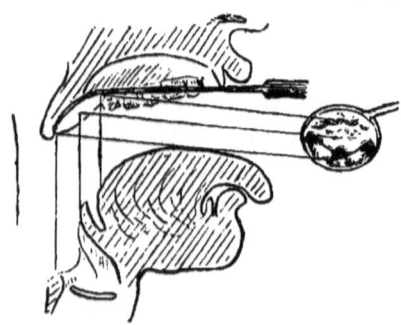

FIG. 7.
Diagram to show the faulty position of the laryngoscopic mirror, which will not give a view of the vocal cords.

ror at the wrong angle, and too far forward, so that only the dorsum of the tongue and the anterior surface of the epiglottis is reflected in it (*Fig. 7.*)

By placing the mirror somewhat further back and

less horizontally, as shown in *Fig.* 6, a more complete image is obtained.

(2,) In introducing the mirror clumsily, the fauces may be unnecessarily titillated, gagging and retching being started. This is especially liable to occur if the posterior pharyngeal wall be touched with the mirror.

(3,) Gagging and retching may be due to a hyperæsthetic condition of the pharyngeal mucous membrane, in which case painting or spraying the fauces with a weak solution of cocaine (5 per cent.), or sucking ice may be tried.

(4,) The dorsum of the tongue may rise so much that either the mirror cannot be introduced, or its reflecting surface is out of view. If forcible *protrusion* of the tongue by the patient, or taking a deep inspiration does not overcome this difficulty, the patient should hold his own tongue while the examiner depresses the dorsum with a spatula in the left hand. Do not drag on the tongue, nor press it unduly on the lower incisors. In all cases let the the patient protrude it, and then simply seize and hold it firmly in position.

(5,) The patient may be tongue-tied and protrusion impossible. In this case the frænum should be snipped, or the dorsum must be simply depressed with a spatula.

(6,) If the tonsils are greatly enlarged, and prevent the introduction of the usual mirror, a smaller one should be used.

(7,) As already stated, the natural conformation of the epiglottis varies greatly. It may be so pendulous as to overhang the larynx so that only its anterior surface is seen, the larynx, or all but its posterior margin, being out of sight. There are several ways of overcoming this difficulty. In slighter cases, the act of phonating *ee ! ee !*, or coughing with the mirror *in situ*, may suffice to raise the epiglottis, when the vocal cords may come into view. Failing by this manœuvre, direct the patient to throw his head well back, place the mirror nearer the posterior wall of the pharynx, and somewhat more vertically than usual, the observer's eye being well above the level of the patient's mouth. In a few cases it is only possible to see the vocal cords by raising the epiglottis with a retractor.

The patient may hold the breath from nervousness; a few instructions and a little patience will soon overcome this difficulty.

I have once seen a persistence of the fœtal occlusion of the glottis by a membranous web; it occupied the anterior two-thirds of the glottic aperture. It was observed many years ago in a young child with no suspicion of syphilis, while I was acting as clinical assistant at the Golden Square Throat Hospital. Mackenzie and Poore have recorded similar cases. The epiglottis may present a deep central notch, even amounting to a bifid epiglottis.

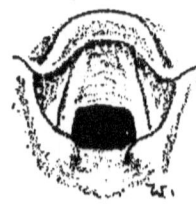

FIG. 8.
Partial occlusion of the glottis by remains of fœtal web.

Trans-illumination of the larynx is sometimes of assistance in noting infiltration of the parts by a growth, or in detecting the consistence of a growth; thus, Mackenzie was enabled to diagnose a small laryngeal tumour as cystic, by this means. The room being completely darkened, a strong light is concentrated in the front of the larynx externally, with the laryngeal mirror *in situ*, sufficient light being transmitted to illumine the larynx.

Chapter III.
ACUTE PHARYNGITIS.

ACUTE CATARRHAL PHARYNGITIS; ACUTE PHLEGMON AND ERYSIPELAS OF THE PHARYNX AND LARYNX; RETRO-PHARYNGEAL ABSCESS.

ACUTE CATARRHAL PHARYNGITIS.

Etiology.—The causes of the affection are :—

(1,) *Idiopathic*—generally due to sudden exposure to changes of temperature, such as coming into the open air from close and over-heated rooms, and is especially prone to occur in damp weather. How far such causes as exposure to cold should be regarded as really microbial in origin it is difficult to say, but recent investigations, such as those of W. H. Park, have shown that staphylococci and streptococci, which are always present in the healthy mouth, increase greatly in number and virulence in damp weather and winter months, and are then capable of setting up acute pharyngitis if applied to the throat.

In this, as in all catarrhal affections of the pharynx and respiratory tract, nasal obstruction is a frequent and potent predisposing factor.

(2,) *Traumatic*—from external violence, or due to injury by a spicule of bone, in swallowing hot fluids, or breathing intensely irritating vapours.

(3,) *Toxic*—dependent on gouty or rheumatic conditions, or the action of various drugs, *e.g.*, antimony, mercury, iodide of potassium, and the symptomatic pharyngitis of various exanthemata, such as measles, small-pox scarlatina, typhoid and typhus fever.

(4,) *Septic*—under which we may include the forms of infection known as hospital sore throat.

The **Symptoms** of course vary greatly according to the severity of the attack, patients often treating themselves by simple and well-known domestic remedies. If due to a chill, the general symptoms of malaise, aching in the limbs, and so forth, will be experienced. In every case soreness of the fauces on speaking or swallowing will be noticed, with a sensation of stiffness in the parts, rendering speech uncomfortable and the voice hoarse and quickly tired. In the earlier stages, the secretions being arrested, the throat feels harsh and dry, and often there is a sensation of a hair or something in the pharynx which cannot be got rid of. After a short time the secretion becomes increased in amount, tenacious and viscid, but is never excessive, thus differing from the condition in acute rhinitis and naso-pharyngitis. The tonsils are usually involved, being red and moderately enlarged, and projecting beyond the pillars of the fauces; often collections of cheesy matter are seen projecting from the follicles, or spreading between the follicles, and occasionally even forming an adherent membrane.

The inflammation is rarely limited to the pharynx, and may extend upwards to the naso-pharynx and nasal passages, or downwards to the larynx and trachea. The uvula too, is often considerably enlarged, and in *acute uvulitis* may become double its size.

We should bear in mind that in diphtheria, scarlet fever or measles, the disease may begin with symptoms indistinguishable from those of simple idiopathic pharyngitis.

Treatment.—In many cases very simple treatment is sufficient to check the attack. Thus, if seen at the onset we may prescribe a hot foot bath with a few tablespoonfuls of mustard on going to bed, or an ordinary hot bath followed by eight or ten grains of Dover's powder. In more acute cases sucking ice is grateful to the patient, and cold compresses may be applied to the front of the neck.

When the trachea is involved and coughing painful, the inhalation of tincture of benzoin, or the same with chloroform, or a mustard poultice applied to the chest will afford relief.

Various lozenges are useful, such as chlorate of potash, cubebs and guaiacum, cocaine alone, or combined with menthol in a pastil.

The bowels should be always freely moved by saline aperients, and food should be light and bland. Internally we may give small doses of quinine or, for gastric catarrh, a bismuth and euonymin mixture. A rheumatic or gouty diathesis is very often the real cause of the condition, and then suitable internal medication becomes necessary.

As the local inflammation subsides the parts should be painted with astringent solutions, such as are used in chronic pharyngitis.

A markedly œdematous uvula may be freely scarified in its lower half, but unless absolutely necessary, the uvula should not be ablated when acutely inflamed.

Any affection in the nose or naso-pharynx, which predisposes to attacks of pharyngitis by causing nasal obstruction, should be attended to.

ACUTE PHLEGMON AND ERYSIPELAS OF THE PHARYNX AND LARYNX.

Etiology.—It is doubtful whether any pathological distinction should be made between erysipelas of the pharynx and larynx, and acute phlegmon.

The streptococcus of Fehleisen has been repeatedly

demonstrated in primary pharyngeal erysipelas, but I believe it has not yet been found in Senator's acute infectious phlegmon. On the other hand Biondi and others have detected Fehleisen's coccus in acute phlegmonous laryngitis. In a discussion on these affections at the Berlin Medical Congress, Semon took the view that acute infectious phlegmon, erysipelas, and angina Ludovici were all due to the same etiological factor, but Fehleisen's inoculation experiments seem to afford strong evidence against this view, and to support Bosworth's opinion, " that while we accept the clinical history as evidencing the fact that phlegmonous pharyngitis is due to a poison of a virulent character . . . it is an entirely distinct affection from erysipelas."

It is unusual for facial erysipelas to spread to the nose or mouth, but it is not very uncommon to meet with cases in which a recurrent facial erysipelas starts from the nose or pharynx and spreads thence to the face, running a fairly mild course. In most of these cases there will be found some diseased condition of the nose or naso-pharynx offering a breach of surface in the mucous membrane, by means of which the infecting coccus gains access to the tissues.

Symptoms.—Acute phlegmonous pharyngitis is characterised by its tendency to extend to the submucous tissues, and, burrowing beneath the cervical fascia, to extend to the trachea, the œsophagus and the tissues of the neck, etc., or to spread rapidly to the larynx, producing fatal dyspnœa from œdema of the glottis. Senator, who has reported four cases in detail, describes the condition as "an acute, febrile disease, beginning with moderate rise of temperature, in which sore throat and pain in swallowing soon set in, followed by laryngeal symptoms, consisting of more or less hoarseness and dyspnœa. Finally consciousness is lost, and death soon follows, without any of the vital viscera being affected. On dissection diffuse purulent infiltration of the deeper parts of the pharyngeal mucosa is found. This condition extends to the larynx and glands, involving secondarily other structures, and thus the clinical symptoms are accounted for."

By extension to the lungs the disease may set up a low form of pneumonia or pulmonary œdema, and death often results from cardiac failure. Senator states that the disease has so far attacked perfectly healthy persons, but Lennox Browne refers to these cases as occurring in persons whose systems have become much reduced by hard work under exceedingly unfavourable sanitary conditions of the inspired air, as well as of the food and water supply in some instances. I have met with only one of these rapidly fatal cases.

J. S., male, age fifty, was perfectly well till 11 a.m. on May 5th, when he noticed pain in the middle of the chest. At 3 p.m. he saw his doctor, who could find nothing much the matter. He had pain and difficulty in swallowing from the first. On admission to the Royal Infirmary he held his neck stiffly, the tissues behind the angle of the jaw and of the hyoid region being somewhat swollen and tender. On examination there was some general hyperæmia of the fauces, pharynx, and larynx, but no œdema, and the œsophageal obstruction was evidently just below the cricoid cartilage. On May 14th five or six ounces of pus were coughed up, and he experienced great relief, but he was so excessively weak that he could hardly be induced to swallow, and in a few hours sank and died. The temperature since admission had fluctuated between 99° and 101°. The post mortem examination revealed the remains of a small sloughy abscess that had formed in the upper part of the œsophagus, but there was little to account for the extremely asthenic and rapidly fatal illness, nor was there anything in his general condition and appearance on admission to lead one to suspect the extreme gravity of the affection he was suffering from.

Erysipelas.—Massei distinguishes two forms : (*a*) those cases in which the general symptoms are most prominent, the local affection taking a relatively mild course ; (*b*,) those in which the local changes are the most marked and urgent features of the case, the general symptoms being less pronounced.

The symptoms of *primary erysipelas of the larynx* are extremely insidious. Massei, in summarising the clinical features states that : (1,) The initial symptoms are, (*a*,) pain in swallowing; (*b*,) fever, which is generally high. (2,) The local symptoms rapidly increase, the dysphagia becoming so great as to result in aphagia, and is soon complicated by dyspnœa from laryngeal implication. (3,) The fever is relapsing in type, a high temperature being followed by defervescence, only to be followed by a fresh rise of temperature. (4,) After a few days usually, the disease terminates in one of three ways : either (*a*,) recovery in consequence of gradual *resolution*, or (*b*,) death from *suffocation*, unless prevented by tracheotomy, or (*c*,) death from *collapse*.

Angina Ludovici is closely allied to acute infectious phlegmon, and is probably due to the same pathological factor. It is a very rare infection, attended by rise of temperature and rapid inflammatory œdema in the chin and front of the neck.

Ludwig's main points in diagnosis are summarised by Parker ("Lancet" Vol. II. 1879): (1,) There may be slight inflammation in the throat which generally disappears in a day or two ; (2,) there is a peculiar *wood-like* induration of the connective tissues which does not pit on pressure ; (3,) this induration spreads uniformly, and is bounded by a well defined border of unaffected tissue; (4,) a hard swelling under the tongue, with a bolster-like swelling along the interior of the lower jaw, of deep red, or bluish colour; (5,) the glands escape, although the disease attacks, or may commence in, the cellular tissue around them.

Ludwig's angina may end in resolution; generally, however, it ends in gangrenous suppuration, and death quickly follows from septic intoxication, or from asphyxia from extension to the larynx.

Treatment in these cases must be prompt and energetic. *Locally*, sucking ice will tend to moderate the intense inflammation, and iced cloths, frequently changed, may be applied around the upper part of the neck. Massei recommends spraying the larynx with perchloride of mercury solution (1 in 2000). A spray or local application of cocaine to the pharynx or larynx will tend to relieve the congestion and pain, and to prevent the occurrence of spasm.

If symptoms of acute dyspnœa appear imminent, the larynx should be freely scarified in the hope that tracheotomy may be avoided.

Bedford Browne has given the details of two remarkable cases in which the free application of sinapisms was followed by immediate relief to the symptoms of laryngeal stenosis. He advises the administration of salicylate of sodium or ammonium.

We must in all cases be prepared to perform tracheotomy at a moment's notice. Very rapid and fatal œdema of the larynx is likely to occur. Intubation has been recommended for these cases, and it has the considerable advantage of obviating the necessity for incising the mucous membrane. The method of performing intubation is described at the end of Chapter XI.

When suppuration has commenced in the larynx

there appears to be less likelihood of the supervention of asphyxia, but the rapid destruction of the deeper tissues may give rise to fatal hæmorrhage. When the acuter symptoms have passed, an antiseptic insufflation may be desirable.

As regards *general treatment*, the indications are to give as much light nutriment as possible, to support the patient with brandy if necessary, and to watch for any symptoms of heart failure. In all cases it is well to give full doses of tinct. ferri perchlor. and to add digitalis and strychnine if the pulse is failing :—

℞ Tinct. ferri perchlor. - - ♏xxv
   Acid. phosp. dil. - - - ♏v
   Tinct. digitalis. - - - ♏v
   Aq. dest. ad. - - - - ℥j

To be taken every four hours.

When the temperature is high, or if rigors occur, ten grains of phenacetine may be ordered, but great care must be observed not to administer antipyretics in doses which will depress the heart, and unless the temperature be over 101°F. it is better to avoid such remedies altogether.

RETRO-PHARYNGEAL ABSCESS.

Suppuration in the cellular tissue of the posterior pharyngeal wall may be primary or secondary.

Etiology and Pathology.—The great majority of cases occur in young children, and are primary or idiopathic, and due to inflammation in the lymphoid tissue of the pharynx. Occasionally it is secondary to a chronic nasal or aural affection left by a former attack of measles or scarlatina. It may be due to caries of the upper cervical vertebræ, or caused by injury from a foreign body.

Symptoms.—The onset may be acute or chronic.

If acute, the symptoms in young children are easily mistaken for those of croup, but in retro-pharyngeal abscess deglutition, as well as respiration, is rendered difficult. The child's cry has a peculiar throaty sound, and respiration may be very much embarrassed, with a croupy cough and dyspnœa.

Objectively a bulging of the posterior pharyngeal wall may sometimes be made out, or at least suspected, and on palpation fluctuation will be felt.

In adults difficulty and pain in deglutition are the chief symptoms, while there is less difficulty in detecting the bulging abscess.

Treatment consists in evacuating the pus, either through the mouth by aspiration or by the knife, or by an incision behind the sterno-mastoid. In one case I found the abscess pointing behind the left sterno-mastoid. When this was opened by a surgeon the retro-pharyngeal abscess became evacuated. When the abscess is opened by the mouth the patient's head should be very low, and the patient turned on his face to prevent the pus escaping into the larynx.

The great danger lies in the occurrence of œdema of the glottis. Ice should be sucked if the patient is old enough, and hot applications made to the neck and submaxillary region. For the treatment of laryngeal complications see Œdema of the Larynx, page 80.

Chapter IV.

CHRONIC PHARYNGITIS.

CHRONIC CATARRHAL, HYPERTROPHIC, AND ATROPHIC PHARYNGITIS; ELONGATED UVULA; VARIX AND ADENOID HYPERTROPHY OF THE TONGUE.

CHRONIC PHARYNGITIS.

THERE is no pathological condition of the pharynx and fauces which is so difficult to define briefly as chronic pharyngitis, indeed the physical signs of the disease are very frequently observed in those who do not complain of discomfort in the throat, and we must recognise that there is no constant relation between the objective signs and the symptoms, and be guided in making a diagnosis of disease chiefly by the subjective symptoms.

For clinical purposes we distinguish three varieties:—

1.—*Simple chronic catarrhal pharyngitis.*
2.—*Chronic granular or hypertrophic pharyngitis.*
3.—*Atrophic pharyngitis or pharyngitis sicca.*

A fourth variety is often described as *exudative pharyngitis*, in which the lymphoid follicles are distended by a collection of cheesy secretion, but this is merely an accidental retention of the secretion and epithelium which has undergone fatty degeneration and which has not, as usual, been extruded.

Etiology.—The causes of chronic pharyngitis are very numerous and diverse. Thus it may be due to:—

1.—Recurrent acute attacks from colds. Hence it is frequently found in a mild form in children suffering from enlarged tonsils and post-nasal adenoids, a condition which may persist in adults.

2.—Nasal obstruction and consequent mouth breathing.

3.—A recent attack of measles, scarlet fever, or influenza, etc.

4.—The gouty or rheumatic diathesis.

5.—The nature of the occupation involving continual exposure to irritating vapours, tobacco dust, stone dust, mattress making, etc.

6.—The excessive use of tobacco, especially tobacco chewing, or free indulgence in alcoholic drinks; it is said also to result from the excessive use of irritating condiments.

7.—Dyspepsia, especially if associated with constipation and portal congestion, is a very common cause.

8.—Improper methods of voice production; wherefore it commonly occurs in clergymen, schoolmasters, and others whose occupation necessitates excessive use of the vocal organs, and who, very often indeed, have had no training in elocution.

9.—The neurotic temperament is an important factor, and accounts for the frequency with which "granular pharyngitis" has to be treated in females.

It is impossible to altogether separate the simple catarrhal and hypertrophic forms either as regards the etiology, symptoms, or physical signs; yet certain broad distinctions may be made. Thus, the first six causes enumerated generally result in a simple catarrhal pharyngitis, the result of more or less acute inflammatory changes. Chronic portal congestion, whether due to gastro-intestinal catarrh, or heart disease, likewise results in a chronic congestion of the pharyngeal mucous membrane. But there are many cases that must be regarded as toxic and due to a sluggish liver failing to arrest and destroy the toxic products of imperfect digestion, which have a specific effect on the pharynx and fauces like muscarin and belladonna. The redness and injection, the dryness of the throat and pain in deglutition, and the sense of roughness or presence of a foreign body in the pharynx, which are characteristic of belladonna or muscarin poisoning, are exactly imitated in dyspeptic pharyngitis, and though these symptoms often occur in a mild degree, their frequent repetition accounts for the permanent alterations in the structure of the mucous membrane. I would explain the pharyngitis due to lithiasis in a similar manner. Many observers, with much reason, regard chronic follicular and simple chronic pharyngitis as altogether distinct affections. Bosworth considers that enlargement of the lymphatic follicles is due, as a rule, to the existence of what he calls the lymphatic temperament, that its origin always dates from early childhood, and that it has no connection with gout or rheumatism ; while Bresgen and others maintain that the enlargement of the follicles is congenital.

The etiology of *atrophic pharyngitis* is an open question. It may apparently result from long standing hypertrophic pharyngitis, while in many cases it arises *ab initio*. It is very frequently associated with atrophic rhinitis, and is usually present in ozæna. We cannot too strongly insist on the state of the general bodily condition as an important factor in the causation of chronic pharyngitis.

Symptoms.—In simple catarrhal pharyngitis there is almost always a sense of weakness and discomfort in the fauces, with constant accumulation of tenacious mucus, which is hawked up with difficulty. The voice is readily tired, and is wanting in resonance owing to the relaxed muscles being hampered by the congested, thickened mucous membrane and accumulation of mucus. When complicated by nasal obstruction, or post-nasal adenoids, the alteration in the quality of the voice is especially marked, the higher notes in the singing voice being the first to suffer. The pharynx is remarkably irritable, gagging and retching being very readily induced, and the uvula is very frequently relaxed and elongated or hypertrophied.

Patients generally suffer from a frequent irritable cough, partly the result of mucus accumulating in the larynx, but largely due to the hyperæsthetic condition of the mucous membrane exciting cough reflexly. A pharyngeal cough, often called a stomach cough, bears no relation to the amount of secretion to be expectorated.

FIG. 9.
Chronic pharyngitis, with elongated and hypertrophied uvula.

On examination, the mucous membrane of the pharynx is found to be diffusely congested, many enlarged veins are seen in the posterior pharyngeal wall, and some hypertrophied lymph follicles are always present.

The *velum palati* and uvula are relaxed and congested, and enlarged mucous glands are found dotted over their surface. The tonsils are generally somewhat enlarged, with gaping crypts; this, however, is not by any means a constant feature. The congested membrane is covered with tenacious mucus, and on examining the larynx, it is generally found to be similarly affected, especially the inter-arytenoid fold.

In *chronic granular pharyngitis*, the discomfort in the throat often amounts to pain of a dragging character, or darting up to the ears. It is especially painful on commencing a meal, and the raw, meaty-looking mucous membrane is readily irritated by hot or stimulating food, and strong wines and spirits. Patients complain of a feeling of a foreign body, a hair or sand in the throat which they cannot get rid of by hawking and coughing.

Occupying the posterior or lateral walls, lenticular, flat, ham-red granules are seen. Saalfeld and Roth* have shown that these granules consist of proliferated adenoid tissue surrounding the orifices of the ducts of the mucous glands.

In the mucous membrane between them, enlarged veins are often found. There is seldom any excessive accumulation of mucus; in fact the mucous membrane is remarkably free from secretion, the patient complaining rather of dryness of the throat, and often there is remarkably little to be seen to account for the severity of the symptoms, and the weakness and alteration in the voice. Thus it is difficult to accept Michel's hypothesis that the toneless character of the voice, and consequent straining, is due to the uneven granules interfering with the resonator functions of the pharynx and naso-pharynx.

In some cases the bands of hypertrophic tissue are

---

* Cited by Schech.

almost confined to the lateral walls of the pharynx behind the posterior pillars of the fauces, corresponding to the salpingo-pharyngeal folds, a variety known as *pharyngitis hypertrophica lateralis.*

Pharyngitis sicca may be regarded as the final stage of chronic granular pharyngitis, when the mucous membrane has undergone atrophy and thinning, both the mucous glands and lymph follicles having partaken in the atrophic process.

In *chronic atrophic pharyngitis*, the thin, glazed, dry mucous membrane allows the colour of the subjacent constrictor muscles to be seen through it. The patient complains mostly of the persistent dryness of the fauces, as well as of the other symptoms mentioned above. When associated with ozæna, a few of the inspissated crusts of fœtid secretion may sometimes be seen behind the soft palate and in the larynx.

Treatment.—Chronic pharyngitis is almost invariably a secondary affection, and therefore when called upon to treat a case, we must always first determine whether it is due to nasal obstruction, or the presence of post-nasal adenoids, or to dyspepsia, heart disease, rheumatism, gout, anæmia, the excessive use of alcohol, and whether chronic constipation is associated with it. It is beyond the scope of this work to indicate the appropriate treatment of the general causes of chronic pharyngitis, while insisting on the necessity for such treatment. The treatment of nasal obstruction from various causes is dealt with in another chapter.

The advisability of recommending a course of saline or aperient waters must depend on the nature of the case. Many chronic cases are wonderfully benefitted by a period of residence at such Spas as Aix-les-Bains, Ems, Homburg, Mont Dore, Bourboule, etc., or one of the health resorts on the northern African coast.

Local treatment is generally called for. The usual astringent lozenges and sprays are very inefficient and disappointing. If the mucus tends to collect in the pharynx and naso-pharynx, a solvent douche, composed of bicarbonate of soda or carbonate of potash (1 to 2 per cent.) with a few grains of boracic acid, may be used once or twice daily. Painting the parts with solutions of mineral astringents, such as sulphate of copper (10 to 20 grs.), nitrate of silver (20 to 50 grs.), sulphate of zinc, or the chloride (10 to 40 grs.), alum (20 to 60 grs.), or the anhydrous persulphate of iron (20 to 60 grs. to the ounce of water) is sometimes effectual. The application should be made daily, first with weak solutions, and less frequently as their strength is increased. An intelligent patient will soon get into the way of painting his own throat, although at first it is necessary for the physician to do it himself till a certain degree of tolerance is established.

A very useful gargle for general use in relaxed throat is a pinch of salt dissolved in a wineglassful of water.

Mandl's solution of iodine in glycerine is strongly advocated. It is used in strengths varying from 6 to 20 grains of iodine to glycerine 1 oz. Begin by using the weaker solution daily, and gradually increasing the strength with less frequent applications. It should be applied to the naso-pharynx and the whole of the pharynx; it causes a burning sensation lasting a few minutes.

Enlarged granular lymph follicles should be destroyed by the galvano-caustic point. A somewhat flat-ended platinum cautery should be used, being placed against the centre of the nodule cold, the current turned on, and withdrawn while still at a bright cherry-red heat, as soon as it begins to burn. Three or four granules may be dealt with at a time. If there are any enlarged veins

CHRONIC PHARYNGITIS.

coursing over the pharyngeal wall, they should be divided in places by the galvano-cautery, so as to obliterate them. Of course such energetic measures require the previous application of 10 per cent. solution of cocaine to the parts to be operated on. This is especially necessary when, as in *pharyngitis hypertrophica lateralis*, the lateral walls of the pharynx require cauterising.

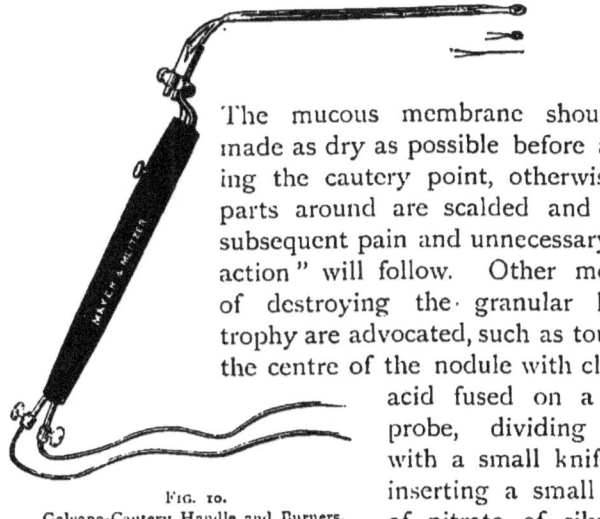

Fig. 10.
Galvano-Cautery Handle and Burners.

The mucous membrane should be made as dry as possible before applying the cautery point, otherwise the parts around are scalded and much subsequent pain and unnecessary "reaction" will follow. Other methods of destroying the granular hypertrophy are advocated, such as touching the centre of the nodule with chromic acid fused on a silver probe, dividing them with a small knife, and inserting a small point of nitrate of silver for a moment, or even curetting. These methods are far less convenient than the cautery. A great deal too much importance has been attached to these small nodules of lymphoid tissue; but I would emphasise the necessity of destroying the enlarged veins which, if left, must, by maintaining the chronic vascular engorgement, render all therapeutic measures futile.

After using the galvano-cautery, the patient should be directed to take only bland, cold food for a day or two, and to avoid smoking. If the operation has been extensive, sucking ice will be a relief; but generally an occasional pastil of morphine and cocaine will be all that is required.

Owing to the constant movements of the constrictor muscles, any operation on the lateral walls of the pharynx is followed by a good deal of pain on swallowing.

A pellicle forms on the points cauterised, and separates in a week or less, leaving a clean surface. The patient should therefore be directed to use some simple saline or cleansing spray or gargle for a few days, subsequent to the cauterising; and to come again in a week's time, when the treatment may be resumed if necessary. Generally several sittings are necessary before a " cure " is effected.

Curetting is sometimes advised for the rapid removal of hypertrophied follicles, while Schech and others have had good results from excising thick and greatly hypertrophied bands, especially in cases where time was an object.

In clergymen, schoolmasters, and professional singers, any faulty method in producing the voice must be detected and corrected. In many cases a few lessons in the elements of elocution will do a great deal for a patient. The greatest difficulty will be encountered with professional singers who have been perfectly trained, whose speaking voice is apparently unaltered in resonance and strength, but who have lost the all essential strength and fulness in their highest notes only. This condition is often associated with a mild form of chronic granular pharyngitis; but while this may require local treatment in the pharynx, we must rely mainly on general measures, rest, and treatment directed to the larynx.

ELONGATED UVULA.

Pathology.—The uvula is almost invariably found to be more or less relaxed in chronic pharyngitis; and whereas the elongation in some cases gives rise to no symptoms, it generally increases the unhealthy condition apart from any of the exciting causes which initiated the mischief, thus often producing alarming symptoms and requiring special treatment. It is necessary however to enter a protest against the habit of some practitioners of snipping off the uvula in all and sundry cases.

Labus of Milan classifies the cases under two heads, viz:—

(1,) In which the uvula and soft palate are merely relaxed, otherwise remaining normal in appearance, there being no congestion or hypertrophy.

(2,) In which there is hypertrophy and chronic congestion of the soft palate and fauces, often resulting in degeneration of the glandular structures of the nasopharyngeal mucous membrane, and associated with severe constitutional disturbance. In these cases a varicose condition of the veins at the back of the tongue frequently results from constant irritation and prolonged congestion.

Symptoms.—In the simpler cases, where there is merely relaxation of the uvula and soft palate without hypertrophy or congestion, the symptoms are mainly impairment of the quality and strength of the voice, and are chiefly observed in professional singers. Labus, as the result of observations in 1,132 cases of professional singers who applied to him for treatment, found that the alteration and impairment of the voice was frequently due, not to the mere elongation of the uvula so much as to the paresis which resulted, preventing the proper and necessary movements of the uvula in singing high notes, and leading to forcing the voice, which becomes flat and tremulous and quickly tired. If the voice be still much used the constant strain may lead to chronic congestion and hypertrophy, and the case passes into the graver type.

In a marked case the patient usually complains of continual hawking, with a feeling of some foreign body in the throat that cannot be coughed up, often likened to a hair or fish-bone in the throat. The cough may be very severe, particularly at night on lying down. The constant titillation at the back of the tongue not infrequently results in vomiting, this especially in the morning or after meals. When the uvula is so elongated as to reach the larynx, laryngeal spasms occur in consequence.

The paresis of the soft palate is partly due to compression of the posterior palatine nerve at its exit from the posterior palatine canal containing the motor nerve fibres for the levator palati and azygos uvulæ, as well as the sensory nerve supply for the velum palati, being analogous to Bell's paralysis, resulting from compression of the facial nerve at its exit from the stylo-mastoid foramen.

In treating cases of relaxed uvula it is well to give local astringent applications a fair trial, especially in the milder cases.

When local applications have failed the uvula ought certainly to be partially removed.

The great amount of benefit that may result from such a simple procedure as removal of the uvula was well illustrated by a case under my care at the Bristol Royal Infirmary. When admitted this case appeared to be far gone in consumption, and, in fact, had been treated for tuberculosis of the lung by his doctor. He was very feeble and emaciated, and *râles* were detected over the whole of both lungs. His uvula was partially ablated, as symptoms pointed to elongated uvula, and the subsequent improvement and final recovery were rapid. He gained 3 lbs. in weight in a fortnight. It is not surprising that the loss of sleep and frequent vomiting should result in great emaciation and weakness, which, in association with cough and expectoration of mucus streaked with blood from the pharynx, might lead one to suspect that the patient is suffering from tubercular disease of the lungs, especially in those who also complain of localised pains in the chest—pains which are purely reflex in origin. The amount of blood that may be lost in this way is very considerable.

Diagnosis.—Cases of elongated uvula, pure and simple, are not always easily recognised. In doubtful cases the patient should be directed to open his mouth, and breathe

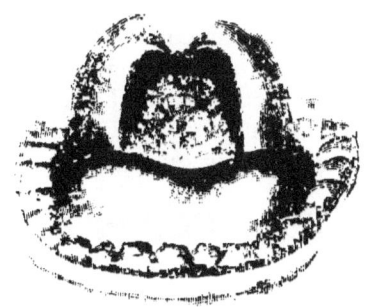

FIG. 11.
Retraction of the normal uvula on striking a high note.

quietly. At first the uvula will be partially retracted into the soft palate, but if elongated it soon drops, and the tip rests on the back of the tongue. On striking a high note the normal uvula is almost completely drawn up into the soft palate which is raised (*vide Fig.* 11), but the relaxed uvula is shortened in wrinkles of redundant mucous membrane (*vide Fig.* 13). The redundant translucent mucous membrane is generally obvious along the free edges of the velum and at the tip of the uvula.

An example of the elongated and hypertrophied uvula,

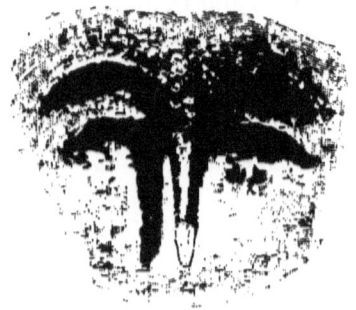

Fig. 12.
Elongated Uvula.

Fig. 13.
The same, on striking a high note, showing the abnormal wrinkling of the redundant mucous membrane.

such as is generally associated with chronic pharyngeal catarrh, is shown in *Fig.* 9, from another patient.

Treatment.—The parts having been cocainised, the tip of the uvula should be seized with forceps and gently drawn forward. The redundant portion is then

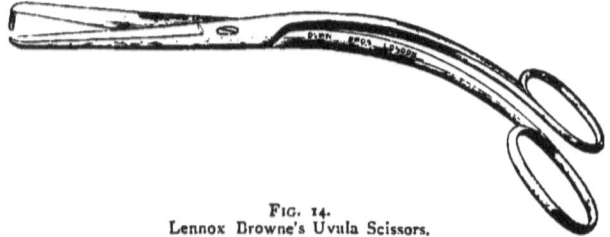

Fig. 14.
Lennox Browne's Uvula Scissors.

removed by one cut with a pair of curved, blunt pointed scissors. By operating in this way the cut surface is situated posteriorly, and is not irritated by food on deglutition.

I have never seen any but good results from the procedure. In one case, however, the son of a medical friend, there was severe secondary hæmorrhage three days after the operation; the possibility of such an occurrence should be borne in mind, and a warning given to the patient or friends. The "gargarisma acidi tannici et gallici" (see Formulæ, No. 15), is useful in case of hæmorrhage.

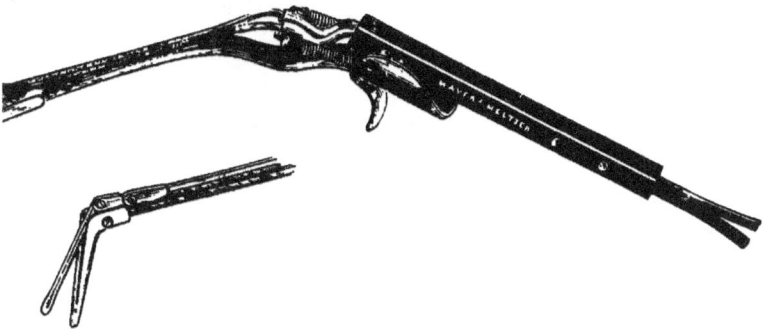

Fig. 15.
MacDonald's Guarded Cautery for the lingual and pharyngeal tonsils.

The food should be soft, bland, and cold, for a few days after the operation, and a mild morphine and cocaine lozenge should be sucked at intervals, especially before meals.

Varix and Adenoid Hypertrophy of the base of the tongue is another condition very frequently observed in association with chronic pharyngitis.

The hypertrophied nodules of adenoid tissue, by impinging on the epiglottis, may cause a constant irritating cough. The most marked case I have met with (as shown in *Fig.* 16), was in a schoolmaster.

Fig. 16.
Hypertrophied adenoid tissue in the glosso-epiglottic fossæ (Laryngoscopic appearance).

These nodules and any varicose veins should be destroyed by the galvano-cautery in the same way as enlarged veins in the pharyngeal mucous membrane.

Chapter V.

DISEASES OF THE TONSILS.

ACUTE TONSILLITIS ; CHRONIC HYPERTROPHY OF THE TONSILS ;
ATROPHIC TONSILLITIS.

TONSILLITIS.

Etiology and Pathology.—A form of tonsillitis is a usual complication of measles and scarlatina, and we may have a diphtheritic or syphilitic tonsillitis. But it is now beyond dispute that there is a close connection between rheumatism and idiopathic tonsillitis. The great majority of cases of acute tonsillitis are truly rheumatic, while the course of the disease and the researches of Hingston-Fox, and Garrod, support the view that tonsillitis and acute rheumatism are microbic in origin. Instances have occurred in my experience which tend to confirm the growing opinion as to the contagiousness of acute follicular, or, more correctly, "lacunar" tonsillitis. Sendziak, of Warsaw, made careful bacteriological examinations (cultures, etc.), in twenty-two cases. He found staphylococci and streptococci in most of them, and in four the pseudo-diphtheritic bacillus; but in none was the Klebs-Löffler bacillus present.

Tonsillitis seems, however, in not a few cases, to be due to drain poisoning, and at times it is impossible to differentiate between some cases of diphtheritic tonsillitis and simple acute tonsillitis from the clinical features alone. It is possible that the few recorded cases with albuminuria and paralytic sequelæ were really diphtheritic.

Tonsillitis is most frequent between the ages of twelve and thirty-five, but it may occur at any age. Reid records a case in a baby aged seven months. Exposure to cold and damp are exciting factors; and tonsillitis is especially prevalent in the early spring and late autumn, being sometimes almost epidemic.

We may distinguish three clinical varieties:—

(1,) *Superficial* or *lacunar tonsillitis*, with diffuse inflammation of the mucous membrane of the tonsil, and accumulation of altered epithelium in the crypts appearing as discrete patches of yellowish exudation.

(2,) *Parenchymatous tonsillitis*, when the deeper tissues of the body are inflamed, the amount of swelling being considerable.

(3,) *Peritonsillitis* in which the connective tissues at the base of the tonsil are chiefly involved.

Suppurative tonsillitis (when pus has formed), is especially prone to follow peritonsillitis, although parenchymatous tonsillitis is also liable to end in suppuration.

Symptoms.—These vary with the degree of inflammation present, generally commencing with soreness and stiffness in the throat for one day, with aching in the back and limbs, followed by a rigor with a rapid rise of temperature, which soon reaches 104° to 105°. The tonsils become greatly swollen, with much pain and tenderness. The tongue is thickly coated, and the bowels constipated.

When the catarrhal inflammation extends to the nasopharynx and the Eustachian tubes, deafness and tinnitus will occur, or the deafness may be due to occlusion of the Eustachian orifices by the swelling of the pharyngeal tonsil. If the high temperature persists, albumen may appear in the urine temporarily.

When suppuration has commenced the pain and tenderness will be greatly increased, with lancinating pains darting up to the ears. In suppurative tonsillitis, usually the second stage of peritonsillitis, the general disturbance and febrile symptoms are often slighter, while the pain is more pronounced than in the first two clinical varieties. Superficial and parenchymatous tonsillitis are generally bilateral, though one tonsil is more affected than the other, while peritonsillitis, the variety that most frequently terminates in suppuration, is often unilateral.

Diagnosis.—It is often a most difficult matter to differentiate between follicular tonsillitis and diphtheria, especially in the absence of any definite membrane in the latter, in which case the chief points in favour of diphtheria are low temperature, little pain, unilateral affection, and the presence of albumen in the urine.

The *prognosis* of simple tonsillitis is always favourable, but we must be on our guard lest, in the earlier stages, we mistake a more serious affection for tonsillitis, especially that rare but very fatal disease, acute infectious phlegmon. Cases of death from suffocation in young children from excessively swollen tonsils are recorded, and tracheotomy has several times had to be performed to avoid asphyxia. Death from rupture of a tonsillar abscess and escape of the pus into the larynx has also occurred.

Treatment.—At the outset, in all cases the bowels should be freely moved by a saline aperient. If the temperature is elevated, the best plan is to give 10 to 20 grains of salicylate of soda every hour, or two hours, till the temperature is reduced, and the pain on swallowing is alleviated. Salicylic acid, salicin, or the soda salt, are universally advocated in cases where there is reason to suspect a rheumatic tendency, but in a great many patients where no rheumatic history, either family or personal, is obtainable, the effect of salicylic acid in abating the symptoms is most gratifying, and in my experience the soda salt is much to be preferred to salicylic acid or salicin.

In young children aconite is an excellent remedy, drop doses of the tincture being given every hour till the temperature is normal.

Guaiacum, first suggested by Sir Thomas Watson, is advocated as a specific in tonsillitis by many authorities. Sajous usually prescribes the ammoniated tincture, one teaspoonful in a half-glassful of milk, in all cases of lacunar and parenchymatous tonsillitis. The patient is first ordered to gargle with a mouthful of the solution, and then swallow it. Enough of the powder to cover a penny is then placed far back on the tongue, the patient being directed to keep it there as long as possible before swallowing it.

Guaiacum lozenges may be sucked in mild cases with advantage. Gargling with dilute solutions of chlorate or permanganate of potash is often grateful, and ice may be sucked.

If the pain and swelling are considerable, gargling may be out of the question, but a spray of cocaine (2 per cent.), or menthol (15 to 20 per cent.), in liquid paraffin tends to diminish pain. Pain on swallowing may be almost annulled by firm pressure immediately in front of the external meatus of the ear on either side—a valuable hint for which I am indebted to Mr. Mark Hovell.

Freely incising the tonsils in one or two places always gives relief; but the rapid diminution of pain and swelling by the free administration of salicylate of soda usually renders such a procedure unnecessary.

Raymond, of Chicago, claims that acute tonsillitis may be aborted by two applications of pure guaiacol to the tonsils. It causes smarting for about five minutes only.

When suppuration has commenced, the inhalation of

steam or gargling with warm water relieves the pain somewhat, and tends to make the pus "point." As soon as there is any indication of the spot where the pus may be found, a fairly deep vertical incision should be made parallel with the free margin of the anterior pillar.

The tonsils should not be removed while inflamed. To this two exceptions may be made : (*a*,) when in children asphyxia is threatened ; (*b*,) in cases of chronic tonsillitis without hypertrophy, which cannot be removed during the periods of quiescence.

CHRONIC ENLARGEMENT OF THE TONSIL.

Etiology and Pathology.—Chronic hypertrophy of the tonsils may be the result of frequently recurring acute attacks.

It is often found in scrofulous children, and is then generally associated with great hypertrophy of the pharyngeal tonsil. Marked hypertrophy of the tonsils is rare after thirty-five, as they generally tend to atrophy after the twenty-fifth year.

We distinguish three varieties :—

(1,) *Chronic lacunar tonsillitis*, attended with accumulation of cheesy matter in the crypts, which gape when the yellow evil smelling masses are extruded.

(2,) *Chronic parenchymatous hyperplasia*, chiefly found in scrofulous children.

(3,) *Chronic fibroid degeneration*, occurring generally in adults, and representing the advanced stage of the hyperplastic form, or associated with the rheumatic diathesis.

Sometimes a pale yellowish swelling may be found, due to occlusion of the orifice of a crypt with retention of the cheesy exudation, one form of the so-called *chronic tonsillar abscess*.

CHRONIC ENLARGEMENT OF THE TONSIL. 49

Symptoms.—With well marked hypertrophy of the tonsils the voice is thick, throaty, and nasal. Respiration may be much interfered with in children, the chest being deformed (pigeon breast), with in-drawing of the ribs corresponding to the insertion of the diaphragm. Snoring and suffocative symptoms during sleep are usual, and in some cases sleep is much disturbed by attacks of alarming spasmodic dyspnœa.

These symptoms of respiratory obstruction are most marked in cases complicated with post-nasal adenoids. Cough is very often present, and the reflex cough set up by enlarged tonsils or lacunar accumulations is not infrequently attributed to other causes.

Some difficulty in deglutition may be noticed, but pain is generally absent, except in subacute attacks of tonsillitis which generally recur at frequent intervals.

Treatment.—The treatment of chronic follicular tonsillitis, with cheesy accumulations and slight enlargement of the tonsil, is often very troublesome. A gargle of chlorate of potash and bicarbonate of soda will sometimes aid in the extrusion of the masses in the follicles. If there are only a few distended follicles the accumulations should be removed by pressure, or by means of a blunt probe, solid nitrate of silver or the galvano-cautery being applied to the empty follicle. Hoffman's plan of breaking down the walls of the follicles with a blunt probe is sometimes desirable. When the tonsil is enlarged it is simpler and more satisfactory to remove it.

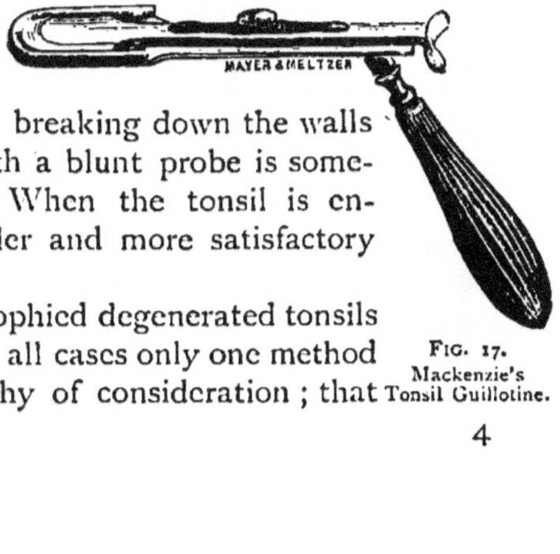

For the hypertrophied degenerated tonsils there is in almost all cases only one method of treatment worthy of consideration; that is, removal.

FIG. 17.
Mackenzie's Tonsil Guillotine.

This is most readily accomplished by the tonsillotome. Mackenzie's tonsillotome is the most generally favoured instrument; Fahnestock's is very largely used, but that devised by Reiner is to be preferred, as it combines the good points in Mackenzie's and Fahnestock's tonsillotomes. A mild antiseptic gargle or spray should be used for a few days after the operation.

Fig. 18.
Reiner's Tonsillotome.

Enucleation of the tonsil is readily accomplished by separating the anterior margin from the anterior pillar of the fauces by the finger nail, and then tearing it out by the tip of the forefinger applied to the superior border of the tonsil; it is an almost bloodless procedure.

Galvano-puncture is largely employed in America for reduction of enlarged tonsils; but it is a tedious and somewhat painful method, and should only be resorted to when a cutting operation is undesirable, as when there is reason to fear hæmorrhage in "bleeders." A fine pointed electrode should be used at a bright red heat, six or seven punctures of a quarter of an inch or more in depth, according to the size of the diseased tonsils, being made at each sitting. The process has to be repeated several times at intervals of three days to a week, till the tonsils are sufficiently reduced in size. The formation of cicatricial tissue completes the reduction. Semon advises (1,) that if the patient is under twenty, and enlargement of the tonsils is transverse, perform the cutting operation; and (2,) that local conditions being the same, but the patient over twenty years of age, give him the option of tonsillotomy or galvano-cautery; but that (3,) if the tonsils are large and productive of serious consequences, and are entirely concealed behind the palatine arches, use the galvano-cautery whatever the age of the patient.

In adults the application of a 10 per cent. solution of cocaine is the only anæsthetic required for tonsillotomy or galvano-puncture; but in young children the operation of tonsillotomy should always be done under nitrous oxide gas or chloroform.

The wire-loop and the galvano-caustic écraseur are still advocated for the removal of the tonsils, mainly on account of the freedom from hæmorrhage ensured by these tedious methods. But dangerous hæmorrhage in tonsillotomy is very rare when proper precautions are observed. Thus Delavan reminds us that Mackenzie, Browne, Schrötter, Gougenheim, Krause, Massei and Capart together have reported about 20,000 tonsillotomies* with but nine cases of hæmorrhage, and of these two were not serious.

De Santi has reviewed the literature bearing on the question of hæmorrhage after tonsillotomy, and arrives at the following conclusions :—

The Causes of Hæmorrhage after tonsillotomy are :— (1,) abnormality in the distribution of the blood-vessels of the tonsils ; (2,) hæmophilia ; (3,) over-use of the voice ; (4,) the too early eating of solids ; (5,) the recorded cases of severe hæmorrhage have occurred very frequently after the use of the bistoury. Nearly all the cases of severe hæmorrhage from tonsillotomy have occurred in adults.

How Should we Treat the Hæmorrhage ?—Most cases tend to cease spontaneously. The patient should be kept quiet and have small pieces of ice to suck. Should the bleeding persist, a mixture of one part gallic acid to

---

* Collected by Désiré.

three parts tannic acid dissolved in water may be sipped, or it may be applied to the bleeding tonsil. If these measures fail, then seek for the bleeding point, and, if possible, it should be seized and twisted with torsion forceps, or it may be necessary to touch the bleeding point or surface with a cautery. Failing these and similar expedients, there remains, as a last resource, ligation of one of the carotid arteries, preferably the external carotid.

Calcareous Concretions in the Tonsil.—These may occur either as small multiple accumulations of calcareous matter occupying the deeper recesses of the crypts, or as large, generally single calculi. Bosworth met with one measuring $1\frac{1}{3}$ inches in length. They consist chiefly of phosphate and carbonate of lime. Grüning has shown that they are originated by the leptothrix buccalis in the tonsillar crypts, much in the same manner as tartar is deposited on the teeth, and the nucleus thus formed accumulates around it, the degenerated mucus, pus and epithelium becoming calcareous. These calcareous concretions may give rise to no symptoms, or they may produce the sensation of an enlarged tonsil, and keep up a chronic inflammation in the affected tonsil, and necessitate removal.

ATROPHY OF THE TONSILS.

Pathology.—Chronic tonsillitis may occur without hypertrophy, two varieties being distinguished by J. Roe : (1,) a chronic disease of the crypts and lacunæ ; (2,) a fibroid degeneration of the stroma of the hypertrophied tonsil, or a cicatricial formation at the base of the tonsil.

Symptoms.—Small hard degenerated tonsils sometimes cause pain, and many of the symptoms of chronic pharyngitis. They are hardly ever found in children, for they only arise in cases of long standing, or chronic degeneration of the tonsils. Walter Dowson has found that degeneration of the tonsils always results in greater or less degree as the result of the tonsillar lesion in scarlatina. A remarkable and very complete atrophy of the tonsils, as well as of the adenoid tissue throughout the naso-pharynx, has been noted by Wingrave in connection with ozæna. Chronic disease of the tonsil without hypertrophy is often overlooked, since more or less atrophy of the tonsil in the adult is a normal process.

Treatment.—When treatment is necessary these tonsils may be drawn forward by a vulsellum, and cut away piecemeal with a blunt pointed bistoury, as recommended by Roe, inasmuch as the tonsillotome cannot be used. Large tonsils should be treated as in ordinary chronic follicular tonsillitis.

The Lingual Tonsil may be the seat of acute inflammation, generally in association with acute faucial tonsillitis, and the inflammation may go on to suppuration. Chronic hypertrophy of the lingual tonsil (*see p.* 43) is commonly found in a mild degree in chronic granular pharyngitis.

Chapter VI.

CHRONIC INFECTIVE DISEASES.

SYPHILIS, TUBERCULOSIS; LUPUS OF THE PHARYNX AND LARYNX, MYCOSIS.

SYPHILIS.

Primary Sore.—Though rare, this has been observed in a good many cases, chiefly on the tonsils; very occasionally on the faucial pillars. "Pain is slight, the tonsil is enlarged and the surface red. Generally there is no marked ulceration; but some erosion is always present, with a well-defined and a sharply-cut margin, although seldom much elevated. The base is generally covered with a slight, whitish, sticky secretion. On palpation there is marked induration, and often stony hardness. The submaxillary glands are enlarged in all cases" (Bulkley Duncan). The nature of the sore is easily mistaken when, as is often the case, it covers the whole tonsil. The well-marked bubo (which seldom or never suppurates), associated with a recent sore throat, should lead to the suspicion of syphilis, which the development of secondary symptoms confirms.

The local treatment consists in prescribing a simple or mercurial gargle.

Secondary Syphilis.

Erythema gives rise to discomfort rather than pain. It presents a peculiar, almost characteristic bright, bluish-red, symmetrical hyperæmia, not extending beyond the soft palate, with an almost sharply defined border. This appearance should always lead to the suspicion of syphilis, evidence of which is generally present in other secondary phenomena.

Mucous Patches.—These bilaterally symmetrical, slightly elevated, bluish-white patches on the tonsils, faucial pillars and pharynx, closely resemble the appearance produced by the application of lunar caustic to mucous membrane, and are surrounded by an erythematous blush. The general symptoms enable one to distinguish this form from diphtheria.

Superficial Ulceration, usually presents the peculiar bilateral "Dutch Garden Symmetry" (J. Hutchinson), so characteristic of secondary syphilitic lesions.

TERTIARY SYPHILIS.

Gummata are rarely seen without ulceration, since syphilitic deposits do not give rise to much pain, and ulceration occurs rapidly. A gumma may arise in the soft palate, in the tonsil, or in the posterior pharyngeal wall, as a smooth or uneven swelling, covered with somewhat congested mucous membrane, which soon shows yellowish spots indicative of commencing necrosis.

Tertiary Syphilitic Ulceration is always due to the disintegration of gummata, and is usually limited by the initial gummy deposit (Bosworth). In early cases these ulcers are found mostly in the soft palate, faucial pillars and uvula. In ulceration occurring many years after the initial lesion they are more often seen on the tonsils and posterior wall of the pharynx as well. While the extent and the depth of the ulcerative process varies greatly, since the amount of gummatous deposit is so variable, we may say that, broadly speaking, the degree of ulceration varies directly as the length of time that has intervened between the primary sore and these later manifestations of the disease.

While both superficial and deep syphilitic ulcers are sufficiently obvious if they occur on the anterior surface of the soft palate and. fauces, etc., the inexperienced observer often fails to detect the ulceration when confined to the posterior surface of the soft palate and uvula, and to the naso-pharynx. An intensely injected boggy looking infiltration of the velum or uvula should put us on our guard, and lead to the inspection of the posterior surface of the soft palate and the naso-pharynx by the rhinoscope. Sometimes an extensive ulceration here extends down to the margin of the velum and can be seen on anterior inspection.

Tertiary syphilitic ulceration is often remarkably rapid, and the tissues affected may be extensively disintegrated before the diseased process can be arrested. The extreme importance of not overlooking the real nature of the case is therefore obvious, since a few days' delay may entail a large perforation of the velum or its complete destruction by an ulceration extending from the posterior surface. Deep syphilitic ulceration is generally followed by distortion and contraction of the tissues involved, and not only is the soft palate often completely destroyed, but the destructive process may have involved the loss of the hard palate and floor of the nasal passages.

FIG. 19.
Cicatricial Adhesion of the uvula to the posterior wall of the pharynx, due to tertiary syphilis.

When the destruction is less extensive the pillars of the fauces are often much contracted, and adhesions between the soft palate and the posterior pharyngeal wall may have produced almost complete occlusion of the naso-pharynx.

Treatment.—(See Syphilis of the Larynx p. 120).

TUBERCULOSIS.

Etiology and Pathology.—Tubercular disease of the pharynx and fauces may be either primary or secondary, but it is almost invariably secondary to tubercular disease of the lung, and the vast majority of cases are acute. Only two or three cases are recorded where the acute form was said to be primary. I have met with one case only; it took the usual form of acute miliary deposit in the soft palate and uvula.

Symptoms.—The acute form usually commences by the onset of pain in the fauces, which are found to be slightly swollen, and at first hyperæmic. The soft palate, if the seat of the deposit, becomes stiff and paretic, and in the course of a day or two several discrete, muddy-gray tubercles are visible, slightly elevated, but obviously beneath the translucent mucous membrane. In from one to three days the tubercles coalesce and begin to ulcerate, fresh miliary tubercles in the meanwhile appearing in the pale mucous membrane, only to pass through similarly rapid phases of development. The paretic, stiffened, soft palate fails to shut off the naso-pharynx on deglutition, even before ulceration has produced extensive disintegration; wherefore the voice is nasal, and fluids run into the nose in drinking. Deglutition is very painful and coughing impossible; consequently the patient is unable to get rid of the copious, sticky, stringy, muco-purulent discharge covering the parts, and can only make feeble attempts at hawking.

As in acute miliary tuberculosis affecting the lungs, there is elevation of temperature, but the emaciation and general prostration are more rapid. In the very rare *chronic form* the ulceration is indolent, and granulations and nodular thickening may cause it to resemble lupus.

The tonsils are occasionally the seat of tubercular deposit, either alone or in association with the fauces. In a case reported by Lublinski, the right tonsil was enlarged and greatly congested, and there were five ulcers on this and two on the left tonsil, varying in size from the head of a pin to a lentil, whose bases were covered with a whitish detritus, the margins of the ulcers being only slightly raised, but somewhat redder than the surrounding parts. The shape of the ulcers was in no way characteristic, the larger ones were somewhat oval.

The **Diagnosis** of the acute form has to be made from diphtheria, follicular tonsillitis, syphilis, herpes and small-pox, while the chronic form must be distinguished from lupus and syphilis.

The **Prognosis** of the acute form is exceedingly grave, the disease almost invariably ending in death in from two to six months. In the chronic form, the disease is more indolent, and may heal under local treatment.

Treatment.—The local treatment of the acute form is mainly palliative and consists in sucking ice, spraying the throat with a 4 per cent. solution of cocaine, or 20 per cent. menthol in liquid vaseline, or the insufflation of powdered boracic acid containing gr. $\frac{1}{6}$ of hydrochlorate of morphine to each insufflation. If not very acute, the daily application of solution of lactic acid (20 to 80 per cent.) to the whole of the ulcerated surface may be beneficial.

(See also treatment of Laryngeal Tuberculosis, p. 131, the same measures being indicated in Chronic Pharyngeal Tuberculosis.)

LUPUS OF THE PHARYNX AND LARYNX.

Lupus of the nose and throat is generally regarded as a rare affection, but it is often present without giving rise to symptoms sufficiently definite to attract the patient's attention.

Middlemas Hunt finds that in a collection of 411 cases of external lupus, no less than 20 per cent. were affected either in the pharynx, larynx, or nose. In 173 cases of lupus of the mucous membranes occurring in the clinic of Doutrelepont,* only 6 were free from cutaneous manifestations of the disease, while of these 173 cases, 75 were affected in the nose, 31 on the palate and uvula, and 13 in the larynx.

It occurs with very much greater frequency in females than in males, and generally reveals itself at, or before, puberty.

Etiology.—As regards the etiology of lupus there is no doubt that the disease is directly due to a specific bacillus; but while the great majority of authorities are agreed in regarding lupus and tuberculosis as one and the same disease manifesting itself under different conditions, their identity is denied by Hutchinson, Kaposi, Macintyre, Campbell and others, and I fail to see that the evidence of their identity is sufficiently strong to out-weigh the clinical evidence of their being distinct diseases.

Symptoms.—Lupus vulgaris develops very slowly and insidiously, for, as already stated, it is often present in the pharynx and larynx without manifesting any definite symptoms. The patients generally complain of some stiffness in the pharynx, or of soreness and tickling sensation, and in some cases of slight dysphagia. If the

---

*Cited by Ramon de la Sota, "Dis. of Ear, Nose, and Throat." Burnett.

larynx be involved, the symptoms are chiefly impairment of the voice or dyspnœa. Pain is generally absent or only slight.

The characteristic aspects of lupus of the pharynx and fauces, are beautifully depicted in the coloured figures that Prof. Chiari, of Vienna, has very kindly allowed me to introduce. When the deposit first manifests itself on the uvula or the free border of the soft palate, we may find localised tumefaction, generally of distinctly heightened colour, less marked, and more limited, than in syphilis or acute pharyngitis, but differing from anæmia premonitory of tubercle; and sometimes the deposit is in apparently healthy mucous membrane. In course of time, smooth, hard nodules appear, varying in size from a pin head to a split pea or larger, of a more distinctly rosy colour. The nodular deposit soon produces a considerable and somewhat characteristic twisted and distorted appearance of the parts, and the uvula is often remarkably elongated as well as thickened.

After a variable period the nodules become soft and apple-jelly like, and ulcerate, or occasionally a nodule is absorbed without ulceration. The ulcers present a serpiginous, worm-eaten appearance, with defined, hard or soft margin, granular and prominent, and with velvety, red, dry, indolent base. The process of ulceration and cicatrisation is very slowly progressive, with periods of increased activity alternating with lengthy periods during which the disease appears stationary. In this manner the whole of the uvula and soft palate may be lost, and the disease may spread to the hard palate. When the tonsil is affected it becomes covered with irregular red nodules and pits of ulceration.

In the larynx the disease usually first attacks the free margin of the epiglottis, which shows unilateral tumefaction, gradually extending to the ary-epiglottic folds and

PLATE II

Lupus of the Palate and Larynx.

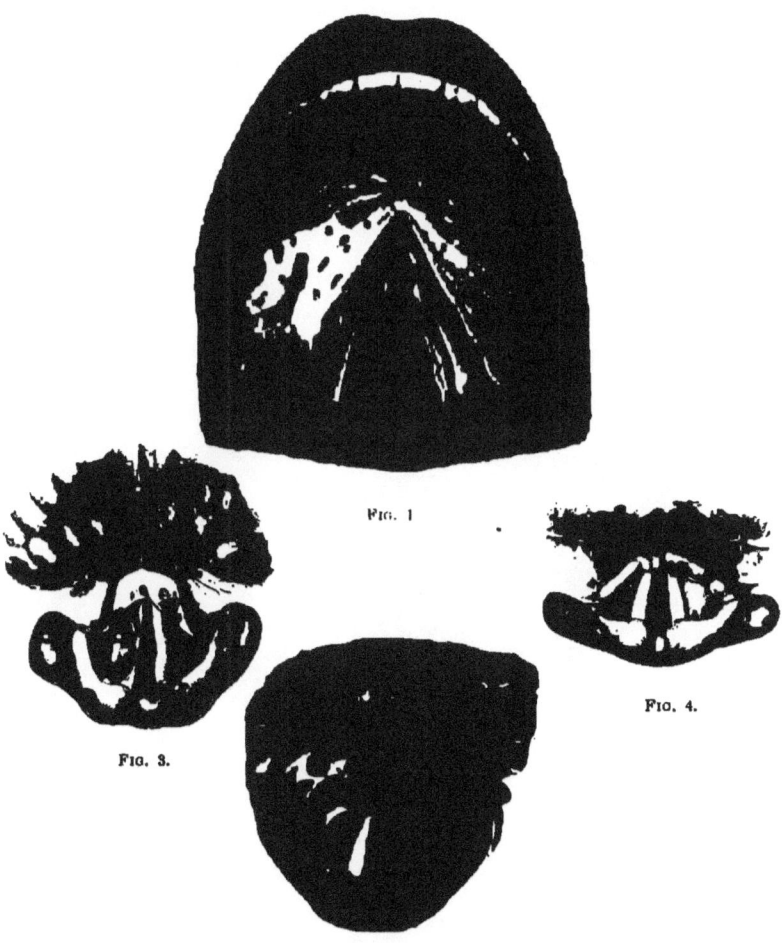

Fig. 1.
Fig. 3.
Fig. 2.
Fig. 4.

Fig. 1.—Lupus of the hard and soft palate and of the fauces, showing cicatrices, disseminated nodules and ridges of tubercles.
Fig. 2.—Lupus of the larynx. At the base of the tongue and on the swollen epiglottis are seen the characteristic tubercles and ulcers; tubercles are present also on the left ventricular band.
Fig 3.—Lupus of the larynx, showing very similar condition. Lupous tubercles on the left vocal cord.
Fig. 4.—Earlier stage of hyperæmia and swelling. The epiglottis partly eaten away by lupous ulceration.

The Author is indebted to Professors O. Chiari and Riehl of Vienna, for permission to reproduce these drawings.

Scott & Ferguson, Edin.

ventricular bands. The vocal cords are the last to be affected; and so slow is the progress of the disease that they often escape, but when they are involved they become red and unevenly tumefied. When the nodules appear on the epiglottis they pass through the same stages as in the palate; but the epiglottic appearance is characteristic, becoming quite pale, worm-eaten and rough, or large portions of it are ulcerated away. The infiltrated ary-epiglottic folds are pale, contracted and tumefied. When cicatrisation has caused a stenosis of the pharynx, dysphagia becomes more marked, and if the soft palate has been eaten away the voice is nasal, and fluids find their way into the nose in deglutition.

When the larynx is first attacked, there may be no symptom to attract attention; but when the disease has extended to the arytenoids and posterior commissure, the voice becomes impaired or lost, while the contraction of the tissues, together with the thickening and the nodular deposits, may occasion such a degree of stenosis that tracheotomy has to be performed. There is never any inflammatory exudation in lupus, and never perichondritis or necrosis of laryngeal cartilages.

Diagnosis.—The diagnosis has to be made from simple chronic pharyngitis, syphilis, cancer, and tuberculosis.

In the absence of cutaneous lupus, the difficulty in excluding syphilis, acquired or hereditary, is considerable, and we are often obliged to wait for the result of anti-syphilitic treatment to prove negative. I have alluded to the main distinctions between lupus and syphilis; and the fact that lupus occurs in the very young, is very slowly progressive, always with cicatrisation, and is almost never painful, together with the characteristic appearances of the growth, and the absence of wasting, fever, or quickened pulse, should rarely leave us in doubt as to the diagnosis. (See also p. 67.)

The **Prognosis** as regards life is most favourable, the only real danger being stenosis of the larynx; but that comes on gradually, and is not very liable to be suddenly increased by perichondritis or œdema, so that tracheotomy can almost always be performed in good time. Haslund records one death from asphyxia.

Treatment.—As regards local treatment, the nodules and tumefactions should be scarified or curetted, and strong lactic acid rubbed in by the methods employed in laryngeal tubercular disease (p. 132). This should be done once a week, successive portions being treated until the whole of the diseased area has become cicatrised and no nodules or ulcers are visible. The case must be watched for a year at least after apparent cure has been effected, and any fresh manifestations must be similarly dealt with. If only tumefaction is present, linear scarification, followed by the application of lactic acid and glycerine, equal parts, or two to one, is recommended by De la Sota.

Isolated deposits may be destroyed by the galvano-cautery. Mandl's solution of iodine, solutions of nitrate of silver, chromic acid solution, 1 in 1,000 solution of perchloride of mercury, are also recommended as local applications.

Stenosis of the larynx may be arrested for a time by intubation, or by dilatation with Schrötter's bougies.

MYCOSIS.

Etiology and Pathology.—The leptothrix fungus is universally present in tartar, decayed teeth, and in the crypts of the tonsils. Under certain conditions it takes root and grows in the mucous membrane, and constitutes the affection *pharyngo-mycosis leptothricalis.*

Symptoms.—It occurs in two forms, the diffuse and circumscribed. In the diffuse form, milky patches are seen usually on the dorsum of the tongue. The more circumscribed form most frequently arises in the tonsillar crypts, from which firmly adherent chalk-white horny excrescences are seen to project. They are also seen at the base of the tongue, on the fauces and uvula, and on the posterior pharyngeal wall.

The symptoms are always either very slight, or altogether absent. Patients may complain of discomfort, stiffness, and dryness in the throat, and the growth may weaken and impair the voice.

Diagnosis.—The fact that the surrounding mucous membrane is healthy, the growths firmly adherent and white, and that there is a complete absence of pain or constitutional disturbance, should prevent any error in diagnosis, while a microscopical examination will show the characteristic thread-like cryptogam in the midst of the amorphous granular matter.

Treatment.—The eradication of the fungous growth often gives considerable trouble. Scraping and the use of the galvano-cautery have generally proved successful when all other measures have failed. When the tonsil is the seat of the growth, Heryng had good results with the galvano-cautery after excision of the tonsil.

Solutions of bichloride of mercury (1 in 1,000), applied locally, in combination with a gargle of the same (1 in 2,000), are favourably commended (Chiari).

Simple removal of the growth is useless, as it invariably grows afresh.

Chapter VII.

NEOPLASMS OF THE PHARYNX AND FAUCES.

BENIGN NEOPLASMS ; MALIGNANT NEOPLASMS ; DIFFERENTIAL DIAGNOSIS OF SYPHILITIC, TUBERCULAR, LUPOUS, AND MALIGNANT ULCERS.

BENIGN NEOPLASMS.

Papilloma is by far the most common form of benign tumour in this region, the small warty growths being attached to the margin of the soft palate, the pillars of the fauces, or to the uvula. They often give rise to no symptoms, unless they attain considerable dimensions. *Adenoma* occurs as a hard growth in the palate or tonsil; *fibroma* and *angioma* are very rare. In fact, "almost every kind and sort of tumour have been observed in the small space of the palate" (Stephen Paget). A small warty epithelioma might easily be mistaken for a papilloma, but it would be associated with hyperæmia of the tissues around.

Calcareous concretions occur in the tonsil, and rarely in the soft palate, and may simulate a growth.

Tr.atment.—A papilloma should be cut off, the tissues immediately around the base being included in the excised growth. The other benign growths should be removed only when their presence causes inconvenience or pain.

MALIGNANT NEOPLASMS.

Symptoms.—*Primary epithelioma* more frequently occurs in the soft palate or pillars of the fauces than in the tonsil. to which it generally spreads, so that it is often impossible to say for certain which was the seat of origin. It presents an uneven surface which soon ulcerates.

In *carcinoma* the most characteristic early symptom is constant pain of gradual onset, increased but not induced by swallowing, of a lancinating character, and darting up to the ear. On digital exploration the enlarged tonsil often gives a characteristic fixed indurated feeling. The neighbouring glands are generally enlarged and indurated; but this condition is by no means invariable, and the absence of obvious enlargement of the glands of the neck does not exclude the diagnosis of cancer. The tonsil soon ulcerates, and then in appearance closely resembles tertiary syphilitic disease. The diagnosis can hardly be made between these affections, till the failure to check the progressive ulceration by anti-syphilitic remedies leaves no doubt as to its real nature. Large doses of iodide of potassium will sometimes relieve the pain, but the malignant ulceration progresses unchecked.

Sarcoma of the fauces and tonsils often develops very insidiously. When, however, the growth has attained any size, the mucous membrane covering it is succulent and bright red in appearance. It is less hard than epithelioma, and varies very much in its rate of growth, some cases showing a tendency to remain localised for a considerable period. It spreads by extension to the neighbouring regions, and very generally involves the deeper tissues behind the angle of the jaw, so as to cause large swellings in the neck. Pain is not a very prominent symptom until ulceration has occurred, and even then is often inconsiderable. Interference with swallowing is chiefly mechanical. Hæmorrhage is often severe, but usually ulceration is not deep.

In carcinoma the onset is equally insidious; but it

develops more rapidly than sarcoma, pain is more marked, ulceration occurs earlier, and the growth is hard, pale pink, or even bluish-pink, with a well-marked areola round the growth. The ulcers are deep, with hard raised margins, and are covered with yellowish-gray *débris*. Carcinoma of the tonsil occurs generally after the fortieth year. While enlarged tonsils dating from childhood may persist throughout life, an enlargement commencing in an adult, especially if unilateral, must always be regarded with grave suspicion.

It is very difficult to arrive at definite conclusions as to the relative frequency of the various forms of malignant disease in the tonsils and soft palate, especially as sarcoma and encephaloid cancer are so often confused, their clinical characters bearing a close resemblance.

Prognosis.—As spindle-celled sarcoma of the tonsils grows less rapidly than the round-celled form, occasionally remaining encapsuled for some time, when secondary extension is slow to appear we are justified in giving a *relatively* hopeful prognosis in the former class if they come under operation early. One of the cases related by Newman* well exemplifies this point. A lady aged fifty-seven had an encapsuled spindle-celled sarcoma of the left tonsil, of slow growth, the lymphatic glands not being involved. He operated through the mouth. After five years there was no local recurrence, but a second sarcoma had formed in the right tonsil with rapid involvement of the lymphatic glands, palate, and pharynx. Three months later she died from hæmorrhage and exhaustion. The second and rapidly fatal growth was a round-celled sarcoma.

These remarks may, perhaps, appear unduly favour-

---

* "Malignant Disease of the Throat and Nose."

able, for it is hardly necessary to say that any form of malignant growth, especially in a region so difficult of access as the tonsil or fauces, is peculiarly grave; but, on the other hand, there is too great a tendency to overlook the lesson taught by such cases as Newman's.

Of course, in every case the diagnosis should be confirmed or refuted by a microscopical examination of portions of the growth.

Treatment.—Malignant growths of the fauces and tonsils should be removed by the knife or cautery, if possible. The choice lies between lateral pharyngotomy and the operation from within the mouth. It is beyond the scope of this work to discuss the relative merits of these operations, or even to indicate the cases in which operation may be successful, but I would emphasise the importance of not regarding malignant disease of the tonsil and fauces, or of the larynx, as invariably hopeless. A large number of recorded cases proves that a radical operation may often be completely successful, if valuable time is not frittered away.

DIFFERENTIAL DIAGNOSIS OF SYPHILITIC, TUBERCULAR, LUPOUS AND MALIGNANT ULCERS.

Syphilitic, tubercular, lupous and malignant ulceration in the nose, pharynx, and larynx present certain characteristics which, without being absolutely pathognomonic, are generally sufficiently definite to enable the practised eye to make a diagnosis in any given case. It is necessary, however, to bear in mind that mistakes are occasionally made by the most experienced, and that we can never afford to dispense with the assistance of any facts in the family or personal history of a case which may aid us in making a diagnosis.

The Superficial Syphilitic Ulcer is definite, cup-shaped or with a flat base, with an even, slightly raised margin, surrounded by a well defined border of brightly injected mucous membrane. The floor of the ulcer is covered with whitish-yellow, sticky, disintegrating muco-purulent *débris*.

The Deep Syphilitic Ulcer is crater-like, with an undermined, slightly elevated, regular, sharply cut margin, surrounded by a well defined areola, with the base covered by yellowish ropy muco-pus and necrotic tissue. The ulceration advances more in depth than in superficial extent. It is followed by cicatricial contraction which gives rise to great deformity.

Tubercular Ulcers present an uneven, ragged, "mouse-nibbled" margin, which is not elevated, surrounded by pale grayish mucous membrane. Ulceration extends superficially rather than deeply, the base being nearly flush with the surrounding swollen tissues; and being covered with grayish disintegrating tissue, it is often difficult to determine the exact limits of the ulcer, which progresses very slowly. In the earlier stages, the miliary tubercles, which have not broken down and coalesced to form the irregular ulcer, may frequently be seen.

Lupous Ulceration is invariably associated with the characteristic nodules, and spreads very slowly in one part while cicatrising in other directions, so that the active ulceration is rarely very extensive at any one time, whilst the mucous membrane surrounding the diseased process is, except for the venous injection, normal in hue. The affected area is always deep red or rose coloured, and the margin of the ulcer is either hard or soft, irregular, defined, and elevated. If ulceration has gone on sufficiently long, the characteristic cicatricial bands traverse the area previously occupied by the slowly advancing ulceration, yet without producing the marked distortion of old syphilitic ulceration. Recurrent ulceration of the cicatrix is pathognomonic of lupus

(Chiari). The floor of the ulcer is depressed and crater-like, velvety and uneven in contour, and seldom presenting any *débris*, but is indolent and dry.

Malignant Ulceration is associated with a definite, and generally increasing tumour, covered with bluish-pink or pale mucous membrane, and surrounded by diffuse injection and infiltration. The ulcer is deep, with well marked, slightly elevated nodular and rapidly advancing margin, the floor of the ulcer being irregular, thickly covered with the foul-smelling disintegrating tissues and muco-pus. Epithelioma often presents bluish-pink, indolent, cauliflower excrescences.

The subjoined table (*vide page* 70) of differential diagnostic signs (from an article by the author in the "Medical Annual," 1894) briefly summarises the main points of distinction between some of the diseases of the tonsils, etc.

CARCINOMA.	SARCOMA.	CHANCRE.
Symptoms. — Dysphagia is always an early symptom, and pain is considerable and persistent, but of gradual onset. Increased pain on swallowing becomes so great as to prevent the patient taking food. Saliva accumulates in the mouth. Early and well marked cachexia, and rapid loss of flesh. *Physical Signs.*—Carcinoma always presents an enlargement with superficial irregularity of surface, which is light pink or bluish, and soon ulcerates with granular fissured surface, hard elevated margin, general cartilaginous hardness and fixedness. Ulceration not very depressed, covered with fœtid mucopus. Early infiltration of neighbouring glands. Hæmorrhage frequent and often profuse, sometimes fatal. Generally unilateral.	*Symptoms.* — Difficulty and pain in deglutition, sometimes very slight, and until ulceration occurs, is chiefly mechanical. Saliva accumulates and dribbles from the mouth. Loss of flesh generally rapid. *Physical Signs.*—Sarcoma attains considerable dimensions before ulceration commences. The growth is red, fleshy looking, and soft, surrounded by a well marked bright red areola. Spreads to neighbouring regions and externally to the neck,—especially rapid is the extension of round-celled sarcomata. Hæmorrhage is frequent and sometimes fatal. Generally unilateral.	*Functional Symptoms.*—The first symptom is a stinging pain in the tonsil, but with little pain on swallowing, which is never so difficult as in cancer or in tertiary syphilis. Cancer occurs in late middle life, but sarcoma may also occur in the young; chancre generally in young adults. *Physical Signs.* — The surface is very red, but there is always a well defined erosion, with sharply cut margin, from the commencement. Induration or even stony hardness. The submaxillary glands early enlarged. Like cancer and tertiary syphilis, and unlike secondary, it is unilateral. No hæmorrhage, only streaks of blood. No emaciation, early appearance of secondary rash, etc. Most amenable to treatment.
SYPHILIS.	TUBERCULAR ULCERATION.	ACUTE TONSILLITIS.
Symptoms. — Pain on swallowing often difficult, never impossible. Wasting and cachexia in proportion to the difficulty in taking nourishment, and not very pronounced. No salivation. In secondary syphilis of the tonsils and fauces there is generally bilateral deposit of mucous patches and superficial ulceration, with well marked purplish areola. In tertiary syphilis the tonsils are irregularly affected by a deep perforating ulcer. The margins of the ulcer are often undermined and overhang the deep lying ulcer, the floor of which is covered with necrotic tissue. The sympathetic glandular enlargement is slight, and not painful as in cancer. Hæmorrhage slight or absent. The rapid improvement under antisyphilitic remedies is always a valuable sign.	*Symptoms.*—Swallowing is always very painful, and loss of flesh rapid, with nocturnal rise of temperature, and a general well marked tubercular cachexia is always present. There is early and rapid infiltration of the parts around, with very early tendency for fluids to return through the nose on swallowing. *Physical Signs.*—General pallor, with diffuse infiltration of the affected region, Early superficial, irregular mouse-nibbled ulceration, with gray débris. In the earlier stages the deposits of miliary tubercles are very characteristic; these ulcerate and coalesce. No inflammatory areola, no sympathetic glandular enlargement. Hæmorrhage generally absent.	*Functional Symptoms.*—Pain very marked from the commencement, great tenderness and difficulty in swallowing. Generally some rise in temperature. In cancer, onset of pain is gradual. *Physical Signs.*—Characteristic redness and inflammatory infiltration. Follicular exudation, but no ulceration. May proceed to suppuration. *Chronic abscess* of the tonsil may be diagnosed by incision and discharge of pus. Most amenable to treatment.

CHAPTER VIII.

NEUROSES OF THE PHARYNX.

SENSORY NEUROSES; MOTOR NEUROSES.

SENSORY NEUROSES.

THE fauces and soft palate derive their sensory nerves from the second division of the fifth, and from the vagus and glosso-pharyngeal nerves. The pharynx is supplied by the glosso-pharyngeal.

Anæsthesia, partial or complete, may be *bilateral, e.g.*, after diphtheria, or in hysteria, or bulbar paralysis; or *unilateral, e.g.*, in the pharynx, from interference with one glosso-pharyngeal nerve by pressure of a tumour.

Hyperæsthesia and *Paræsthesiæ* are often met with in neurotic patients apart from any organic disease. They are also suggestive of gout, or may be premonitory of pulmonary tuberculosis.

Treatment.—Nervine tonics are indicated, and the treatment of any underlying dyscrasia, such as anæmia, gout, dyspepsia, and portal congestion, and any local cause of irritation, should be removed.

MOTOR NEUROSES.

The soft palate and uvula, the levator palati and the pharyngeal constrictors are innervated by spinal accessory fibres in the pharyngeal plexus (Horsley & Beevor), while the tensor palati is supplied by the fifth nerve. Thus paralysis of the soft palate results from cerebral disease in which the spinal accessory is involved, as for instance in bulbar paralysis; or from peripheral nerve lesions, as in diphtheritic neuritis. The view that paralysis of the soft palate is due to, and accompanies paralysis of the facial nerve is no longer tenable.

The accessory nerves may be involved either unilaterally or bilaterally. When the lesion is unilateral, the uvula is drawn towards the healthy side, and the velum is lower and less arched on the affected side. If bilateral, the velum hangs loosely, and does not become elevated when stimulated. The voice is nasal, and fluids may escape into the nose during deglutition. As the paralytic condition of the pharyngeal constrictors becomes more marked, deglutition gets more and more difficult, and the difficulty in swallowing fluids is always greater than for solids. When the difficulty in swallowing is due to obstruction, it is naturally first noticed, and always more pronounced, in reference to solid food.

On the other hand we may have a paralysis of the tensor palati without accessory paralysis.

Spasm of the pharynx is always a functional disorder, generally found in nervous, hysterical patients, and interfering with deglutition; it may occur in association with acute inflammatory affections, *e.g.*, acute tonsillitis. Clonic spasm of the levator palati gives rise to a peculiar sound audible to the patient and those around. The cause is obscure; but it is regarded by many observers as a reflex neurosis, and therefore any possible source of irritation in the nose or naso-pharynx should be sought for and rectified.

Treatment.—In addition to general treatment, appropriate to the condition with which these motor neuroses are associated, the application of the faradic, or galvanic current will often prove most efficacious in relieving spasm.

Chapter IX.
ACUTE INFLAMMATIONS OF THE LARYNX.

ACUTE CATARRHAL LARYNGITIS; SPASMODIC LARYNGITIS; ŒDEMA; PERICHONDRITIS OF THE LARYNGEAL CARTILAGES.

ACUTE LARYNGITIS.

ACUTE catarrhal laryngitis in adults is a minor ailment that is sufficiently familiar to most people, and, except in professional voice users, often hardly attracts the serious notice of the patient. In children, however, owing to the anatomical peculiarities, it is frequently attended with the gravest symptoms.

Etiology.—The usual *exciting* cause is exposure to damp, cold, or any sudden change in temperature; or it may be set up by inhaling steam or irritating vapours, etc., by the action of dust, or excessive straining of the voice. *Predisposing* causes are sedentary habits, the rheumatic or gouty diatheses, and all forms of nasal obstruction producing mouth-breathing. It may occur during an attack of measles, typhoid fever, and more rarely in scarlet fever, and typhus. Frequent attacks of laryngitis often precede true laryngeal tuberculosis, and should lead one to examine the lungs for any indications of commencing phthisis.

Symptoms.—*In adults* the symptoms are purely local, or associated with those of a severe cold. There is hoarseness, or more or less complete aphonia, and a dry, tickling cough, with little or no expectoration, unless the inflammation extends to the trachea and bronchi.

The laryngeal mucous membrane is heightened in colour, and this alteration is most striking in the normally yellow epiglottis and white vocal cords. In the milder forms, there is at most only slight thickening of the mucous membrane, and only a very slight secretion of viscid mucus on the vocal cords.

When the attack is more severe the submucosa becomes infiltrated, and the lower part of the epiglottis, the ary-epiglottic folds, and the loose mucous membrane in the inter-arytenoid fold may be obviously thickened. The symptoms are then greatly aggravated, and, in addition to more marked constitutional disturbance, the voice is completely lost, and there may be considerable inspiratory stridor from the resulting laryngeal obstruction, and a loud barking croupy cough. Though it is unusual to find marked symptoms of laryngeal obstruction in adults, the occasional supervention of acute œdema must not be forgotten. Even in the milder forms it is common to find imperfect approximation of the vocal cords on phonation from paresis of the internal thyro-arytenoidei muscles, or of the arytenoideus. Gerhardt affirms that the paresis of the thyro-arytenoidei in laryngitis is due to implication of the nerve endings, while Schrötter believes that it is due to inflammatory infiltration of the muscles; it is probably due to both these causes.

Local variations in character and distribution of the inflammatory changes have been distinguished by the terms *epiglottitis, arytenoiditis,* or *chorditis* when the epiglottis, arytenoid folds, or the vocal cords are respectively the only parts implicated; *l. hypoglottica,* when the submucosa below the true cords is œdematous; *l. hæmorrhagica,* when the local inflammation is attended with hæmorrhage; *l. herpetica,* when vesicles are present and associated with cutaneous or pharyngeal herpes; *l. sicca* is generally a chronic affection, but acute laryngitis without exudation has been described under this name.

In children acute laryngitis is generally productive of severer symptoms than in adults. Sappey has shown that in children the lymphatic supply of the mucous membranes is far more extensive than in adults, and Bosworth cites this fact as a possible explanation of the tendency for acute laryngitis in children to be attended with subglottic infiltration of the submucosa. In childhood, too, the glottic opening is *relatively* (as well as absolutely) small, and the mucous membrane is more vascular and less firmly adherent to the underlying structures. Moreover, in children, paresis of the intrinsic muscles of the larynx is more readily induced by inflammatory infiltration, than in older patients, and there is a greater tendency for reflex nerve phenomena, such as laryngeal spasm, to occur. Thus we have a sufficient explanation of the usually marked clinical difference between acute laryngitis in adults and in children. For these reasons, even simple catarrhal laryngitis in a child must not be regarded too lightly. Occurring in the first two or three years of life it is certainly a dangerous affection; and, as is the case, in all inflammatory affections of the respiratory tract, the danger is more or less in inverse ratio to the age of the patient. The temperature is moderately febrile, $100°$—$101°$ F., the pulse quick and hard, and the respirations frequent and laboured. At first the cough is harsh, soon becoming loud and brassy, and usually there is some laryngeal spasm following the cough. In a short space of time muco-purulent expectoration becomes copious, and the cough is then less distinctly croupy in character. The symptoms are worse towards evening but are not sudden in onset, nor is there the marked intermission in the laryngeal stridor that is always observed in spasmodic laryngitis.

SPASMODIC LARYNGITIS.
(LARYNGITIS STRIDULOSA OR FALSE CROUP.)

This form of acute laryngitis must not be confused with the purely nervous affection *laryngismus stridulus* on the one hand, and true membranous croup on the other. In spasmodic laryngitis we have the symptoms of catarrhal laryngitis with more marked laryngeal spasm and less catarrh.

The usual history is that the child has caught a cold, and respiration is a little embarrassed towards evening. There is a dry hard cough, which, however, does not prevent it falling asleep. Towards midnight it suddenly awakes with a laryngeal spasm; respiration is greatly embarrassed, the cough is loud and brassy, with marked inspiratory stridor from laryngeal obstruction, which persists till death seems imminent from acute asphyxia. In a few minutes the spasm passes off, respiration becomes easier, and the child falls asleep; and though restless and disturbed by occasional croupy cough, the spasm may not recur again. Towards morning there is marked remission of the symptoms; but the following night or two the laryngeal spasm recurs, yet, almost invariably, with diminishing severity, the catarrhal symptoms and croupy cough persisting for some days.

One attack predisposes to others, but this is largely due to the fact that spasmodic laryngitis is generally associated with what Bosworth terms "lymphatism;" for in children subject to false croup we often find other evidence of this in the presence of enlarged tonsils and post-nasal adenoids. This authority distinguishes spasmodic laryngitis by the term, acute subglottic laryngitis (*laryngitis hypoglottica*), since the larynx generally exhibits characteristic subglottic infiltration of the mucous membrane.

(Contrast the symptoms with those of true croup, p. 91.)

Treatment.—In adults, simple acute laryngitis should be treated by absolute rest of the voice and confinement to a warm room. Free evacuation of the bowels, and abstinence from the use of tobacco and alcohol, should be enjoined. A cold compress may be applied externally to the laryngeal region, till all the acute symptoms have passed off. Ice may be sucked, or we may resort to steam inhalations containing compound tincture of benzoin, camphor, eucalyptus oil or creasote; or an atomiser, containing menthol with, or without, a small quantity of camphor, cocaine, or morphine, may be sprayed into the the larynx several times a day. Internally ipecacuanha and cubebs, or ammonium chloride, are of service.

When the acuter symptoms have passed off, more stimulating inhalations are often desirable, such as oil of Scotch pine, or the *pinus pumilio*, ʒj to the pint of nearly boiling water, and the steam inhaled.

Professional singers and public speakers frequently require that, if possible, an attack of acute laryngitis should be aborted, or at least kept in abeyance sufficiently to enable them to keep some important engagement. In such cases the bowels should be freely evacuated by some saline aperient or calomel. Absolute rest of the voice, a mustard poultice externally, and sucking ice at frequent intervals may be ordered. A pastil containing morphine (gr. $\frac{1}{12}$), cubebs (gr. $\frac{1}{2}$), atropin (gr. $\frac{1}{120}$), proto-iodide of mercury (gr. $\frac{1}{60}$), may be given every four hours, till six have been taken. If only a few hours can be given for treatment, a hypodermic injection of gr. $\frac{1}{30}$ of strychnine may be given. Twenty minutes before the engagement a glass of vin Mariani de coca should be taken, and the following solution be used with an atomiser and well inhaled :—

℞ Menthol	- - -	grs. xx
Morphine	- - -	gr. ¼
Cocaine	- - -	grs. iv
Colourless vaseline oil		℥ss

In young children there is almost invariably some febrile disturbance, and minute doses of aconite frequently repeated should be administered to reduce the temperature. Confinement to a warm, properly ventilated room is obviously essential; and if there is considerable respiratory embarrassment, the patient should be placed in a steam bed, and hot moist applications kept on the throat. Ipecacuanha, apomorphine, or tartar emetic should be given frequently in small doses. When there is difficulty in coughing up the viscid expectoration, the early administration of a non-depressing emetic, such as sulphate of zinc or copper, or of ipecacuanha wine, often affords marked relief. Free evacuation of the bowels is important in children as in adults.

If laryngeal obstruction persists, it may be necessary to perform intubation, or tracheotomy. To my mind there is no doubt that intubation should be preferred if medical aid can be summoned within half an hour, should the tube be coughed up. The great danger, next to acute asphyxia, is the occurrence of wide-spread pulmonary engorgement and catarrhal pneumonia, and, since this is certain to come on should the urgent dyspnœa be allowed to continue, it is better to intubate quite early, for in non-membranous laryngitis, and in skilled hands, the procedure is practically unattended with danger. (See Intubation, p. 101.)

After recovery from the acute symptoms great care must be exercised in preventing a recurrence of the attacks, and exercise in the fresh air, plain food and warm clothing, will be necessary

ŒDEMA OF THE LARYNX.

ACUTE INFLAMMATORY ŒDEMA.

Etiology.—Inflammatory œdema *(See Plate III, Fig. I.)* rarely supervenes on an acute catarrhal laryngitis. It is sometimes more or less localised as an *epiglottitis* (miasmatic), *arytenoiditis*, or *chorditis*. *Laryngitis hypoglottica* should be regarded as an acute subglottic œdema. Inflammatory œdema may occur in scarlet fever, typhoid fever, and small-pox; and Pugin Thornton has seen three cases following influenza. Acute infectious phlegmon of the larynx may arise by an extension of acute phlegmonous pharyngitis, or this very fatal disease may be primary in the larynx (see p. 23). Most of the cases of idiopathic abscess of the larynx are probably of this nature. Semon states that "idiopathic primary, acute œdema of the larynx is excessively rare. According to Sestier's excellent statistics on œdema of the larynx, simple inflammation was the cause of œdema in only 6 per cent. of all his cases; and Dr. Morell Mackenzie believes that in nearly all these instances of so-called simple inflammation, the disease is due to blood-poisoning (*Lancet*, April 1st, 1882)."

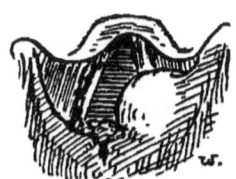

FIG. 20.
Perichondritis and secondary œdema of the left arytenoid and ary-epiglottic fold in a tubercular larynx.

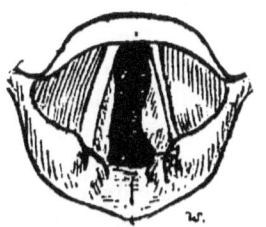

FIG. 21.
Laryngoscopic appearance in subglottic œdema.

Acute secondary œdema may occur in syphilitic, cancerous, or tubercular (rare) disease of the larynx.

Symptoms.—They are those of catarrhal laryngitis, with more pain, dyspnœa and greater constitutional disturbances super-added. When the epiglottis is implicated, there is difficulty in deglutition; and, if the arytenoid folds and ventricular bands are involved, respiration will be embarrassed from the resulting laryngeal stenosis, more especially if the sub-glottic tissues are implicated. My drawing (*Fig.* 21) shows a wider glottic aperture than was actually present in the case it was taken from.

Objectively the epiglottis is seen as a pink sausage-shaped swelling, and beneath it the swollen rounded arytenoids and ary-epiglottic folds, which lose their normal contour, and conceal the infiltrated ventricular bands and the subjacent vocal cords (see *Plate III., Fig. 1*). When the parts above the vocal cords are not much affected, it may be possible to see the subglottic swelling below the true cords.

Treatment.—*Acute inflammatory œdema* should be treated on the same lines as the severer forms of acute catarrhal laryngitis; but in the more urgent cases leeches should be applied to the larynx externally, followed by the ice-bag, and free scarification of the œdematous swelling should be always performed, if practicable. The swelling then generally rapidly subsides, and though it may recur, and the scarification have to be repeated, the relief is usually very marked. Should it fail to relieve the threatening asphyxia, tracheotomy should not be delayed. The trachea should be opened low down, as subglottic œdema is generally associated with the supraglottic inflammation. Intubation may be tried. In one case I was successful in relieving the patient, an adult, by this means, when almost suffocated, and though the tube had to be replaced once or twice, as it was only a child's tube, the necessity for tracheotomy

PLATE III.

Acute Inflammation—Laryngeal Tuberculosis.

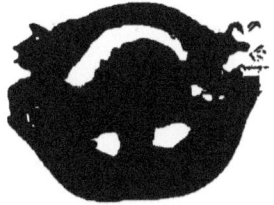

Fig. 1.   Fig. 2.

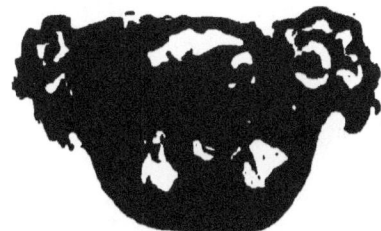

Fig. 3.

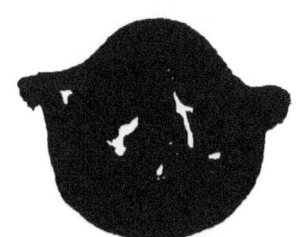

Fig. 4.

Fig. 1.—Acute inflammatory œdema. On the left is seen commencing ulceration.

Fig. 2.—Laryngeal tuberculosis.—The whole of the larynx is anæmic and infiltrated with tubercular deposits, forming characteristic anæmic sausage-shaped swelling of the epiglottis, pear-shaped arytenoids, and swollen ventricular bands.

Fig. 3.—Laryngeal tuberculosis, advanced. The epiglottis is breaking down, and extensive ulceration has occurred on the anterior surface along the lateral glosso-epiglottic folds. In a syphilitic patient.

Fig. 4.—Laryngeal tuberculosis.—The arytenoid regions are diffusely infiltrated. Outgrowths are seen projecting from the arytenoid space and under surface of the left vocal cord. Both cords are ulcerated; the ventricular bands also show irregular infiltration. The patient died within a few months, with both lungs breaking down.

P. Watson Williams,
ad naturam del.

Scott & Ferguson, Edin^r

was obviated. As a matter of fact, the intubation instruments were at hand, and there was no time to get tracheotomy instruments in this case ("Brit. Med. Journ.," 1888).

Scarification may be done in some cases with the aid of a laryngeal mirror. Mackenzie's or Heryng's scarifier may be used; or, failing any special instrument, an ordinary sharp-pointed curved bistoury, with all but the last half-inch of the blade wrapped in lint or plaister. If dyspnœa is urgent, the larynx may be scarified without inspection, guiding the punctures by the left forefinger tip in the larynx.

NON-INFLAMMATORY ŒDEMA.

Etiology.—It occurs in renal disease with or without general anasarca, and very rarely in diabetes and obstructive valvular heart disease. It may be produced by local injury, or by pressure on the veins of the neck by tumours, by scalds, or from the internal administration of iodide of potassium. It commonly supervenes in syphilitic or cancerous affections of the larynx, and, more rarely, in tubercular disease and diphtheria. Strübing describes a form of œdema which he calls angioneurotic, which appears and disappears very rapidly, and is generally associated with a similar condition occurring in the face.

Symptoms.—In non-inflammatory œdema the symptoms are simply those due to laryngeal obstruction, or to mechanical interference with the action of the vocal cords, and vary with the extent and locality of the inflammation. Objectively the parts affected are swollen and translucent, the contour of the laryngeal structures being much the same as in inflammatory œdema.

Treatment.—*Non-inflammatory œdema* may call for relief by scarifying, or require tracheotomy; while appropriate general treatment is indicated in renal and cardiac affections, and in syphilis, etc. Pilocarpine, injected hypodermically, has been used with success.

CHONDRITIS AND PERICHONDRITIS OF THE LARYNX.

Ossification of the laryngeal cartilages generally occurs more or less extensively as old age approaches, and is simply one of the indications of the degenerative changes in the tissues of those advanced in years. It sometimes takes place in middle life, but is without significance.

Fibroid degeneration of the cartilages has been described, but is an extremely rare condition, and one I have never met with.

Perichondritis is a very common complication of advanced malignant, tubercular, and syphilitic disease of the larynx. In association with these affections it is generally subacute, but occasionally leads to suppuration.

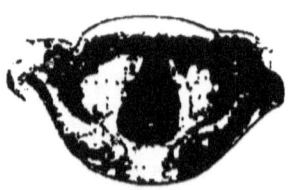

FIG. 22.
Syphilitic perichondritis of the right ala of the thyroid cartilage, giving rise to a swelling beneath the right ventricular band.

FIG. 23.
Perichondritis of the cricoid cartilage.

Etiology and Pathology. — Primary perichondritis is usually acute or subacute, and due to cold or traumatism. Acute perichondritis may follow typhoid fever or phlegmonous laryngitis and erysipelas. Apart from syphilis, tubercle or malignant disease, it most frequently affects the cricoid. If the perichondrium of the inner surface is the seat of the inflammation, an irregular, nodular, unilateral, inflammatory swelling can generally be recognised, encroaching on the subglottic space, and pushing up the vocal cord; or the posterior surface only may be affected.

The thyroid perichondrium may be inflamed internally, forming a smooth red swelling beneath the ventricular band. If the external surface is the seat of the inflammation, tenderness and swelling over the thyroid will be obvious.

The arytenoid is sometimes the cartilage affected, appearing as a red smooth swelling, involving more or less fixation of the corresponding cord.

Symptoms.—Acute perichondritis is often ushered in with chilliness or a rigor; the temperature in such cases is febrile. When the cricoid is involved on its inner surface, considerable dyspnœa and aphonia may result. If the perichondritis is on the posterior surface, pain on deglutition is a prominent feature.

The exudation and swelling may undergo resolution, but, in many cases, suppuration, with necrosis of the affected cartilage, results. This is especially liable to follow cricoid or arytenoid perichondritis. In that case, purulent exudation may persist for months or years until the necrosed sequestrum is exfoliated, during which the patient presents a miserable aspect, and becomes greatly emaciated from the pain and dysphagia and want of sleep.

Treatment consists in the application of ice externally, giving the patient ice to suck, and treating the general symptoms. If dyspnœa is marked, laryngotomy may be required. When suppuration has occurred, with consequent necrosis, the dangers are considerably increased, and many patients succumb to this disease; therefore the strength must be supported by general tonics and dietetic measures. If possible, the necrosed sequestrum should be removed as early as it can be accomplished.

Chapter X.

CHRONIC LARYNGITIS.

CHRONIC LARYNGITIS, AND PACHYDERMIA LARYNGIS;
CHORDITIS TUBEROSA.

CHRONIC LARYNGITIS.

Etiology.—Chronic laryngitis in a large percentage of cases dates from an acute or subacute attack, and thus all causes of acute laryngitis are important etiological factors. It may be chronic from the outset, and due to excessive use of the voice, working in ill-ventilated rooms, or breathing dusty atmospheres, or to chronic dyspepsia, and the abuse of alcohol or tobacco. It is often associated with chronic pharyngitis and naso-pharyngitis, or general anæmia. Persistent laryngitis often precedes tuberculosis of the larynx or lung; it is also predisposed to, or directly caused by, the rheumatic or gouty habit. Undoubtedly a prolific cause of intractable laryngitis lies in nasal obstructions, with resulting mouth-breathing.

Symptoms.—The chief symptoms are huskiness of the voice in varying degree, and vocal weakness. The patients complain that the voice is quickly tired, and becomes hoarse, or goes altogether; there is a sense of dryness or tickling in the larnyx, and, in the hyperplastic forms, slight or considerable obstructive dyspnœa. Cough is often complained of, with little expectoration, unless, as is so often the case, the trachea and bronchi are likewise the subjects of chronic catarrh.

Examination shows in simpler cases that the laryngeal

mucous membrane is somewhat redder than usual, either diffusely or in patches. The vocal cords are slightly pink or gray, and sometimes small vessels may be seen coursing over them. Small accumulations of mucus may be found on the ventricular bands and vocal cords, or lying along their free margins, the sticky mucus stretching across the glottic opening, especially near the anterior commissure. When persistent, the term *Laryngorrhœa* is sometimes applied to these cases. *Fig. 24* is taken from a professional bass. He dated all his trouble from straining the voice in using the "côup de glotte" in shouting to a friend.

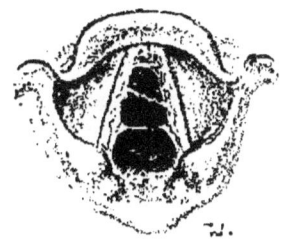

FIG. 24.
Strings of sticky mucus stretching across the glottis in laryngorrhœa.

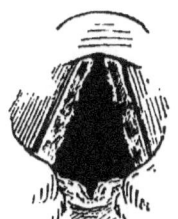

FIG. 25.
Pachydermia diffusa of the vocal cords, and of the inter-arytenoid fold.

In most cases we see other changes as well, giving varieties that have been termed the hyperplastic form, or *pachydermia diffusa*. The most common situation of the hypertrophy is in the posterior commissure, causing wrinkling of the mucous membrane, or mammilliform outgrowths, which may interfere mechanically with the approximation of the cords, or the vocal cords may show nodules on the vocal processes (*pachydermia laryngis*, Virchow). The ventricular bands are often thickened so considerably as to conceal the vocal cords altogether. The subglottic form of hyperplasia, *chorditis vocalis hypertrophica inferior*, shows the infra-glottic swelling below the vocal cords on deep inspiration. A form analogous to chronic follicular pharyngitis is termed *granular laryngitis*.

The *œdematous* form is characterised by diffuse or localised œdematous infiltration, while the *hæmorrhagic* variety is attended with marked hyperæmia followed by hæmorrhage.

Superficial catarrhal ulceration, especially of the vocal cords, may occur, but it should lead to the suspicion of tuberculosis.

Treatment.—Any constitutional condition that may act as a predisposing cause of the affection, *e.g.*, rheumatism, gout, dyspepsia, anæmia and constipation, must of course receive attention; but whatever is the directly exciting cause, any departure from a condition of perfect health must be remedied if we hope to afford lasting relief from such an intractable affection as chronic laryngitis. Any evidence of nasal obstruction should be noted. The voice should be used as little as possible, and all exciting causes, whether lying in the habits or occupation of the patient, should be avoided.

In clergymen, schoolmasters, vocalists, and public speakers, laryngitis is very frequently brought on, not so much by excessive use of the voice as by wrong vocal methods, and the importance of attention to this point cannot of course be over-estimated. Ellis has shown that in schoolmasters the inhalation of chalk dust from the blackboard is a frequent cause of the affection.

As regards the loss of the singing voice, the vocal failure is often due to either: (1,) wrong methods of breathing; (2,) the wrong use of the vocal registers; (3,) the simple over-use of the voice; or (4,) the use of the voice while suffering from catarrh (Hunt).

I have already alluded to the importance of attending to the general health, and a simple tonic treatment, such as the administration of iron and quinine, strychnine and phosphorus will often prove highly beneficial, whilst change of air to a dry bracing climate, fresh air and exercise, a morning cold bath and similar measures, are as important as the administration of drugs.

Of internal "specific" remedies the most useful in my experience are bichromate of potash (gr. $\frac{1}{30}$) or the proto-iodide of mercury (gr. $\frac{1}{16}$) three or four times daily, and cubebs or ammonium chloride. Locally, stimulating inhalations of turpentine, oil of Scotch pine, camphor, creasote, or eucalyptus, used at night; and during the day the oil atomiser of eucalyptol (gr. xx), camphor (gr. i to ij to the ʒj of liquid vaseline), or inhalations of nascent chloride of ammonium vapour, are beneficial.

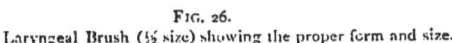

FIG. 26.
Laryngeal Brush (½ size) showing the proper form and size.

As a rule applications to the larynx by the brush should be avoided, unless there are hypertrophic changes. Then such astringents as chloride of zinc (grs. x to xxx to ʒj), or alum (grs. v. to x), or perchloride of iron (grs. ij to xx), may be applied with the laryngeal brush or cotton wool holder. In making these applications the laryngoscopic mirror is held in the left hand, and thus the brush can be guided by sight, the patient himself holding the tongue in a small towel.

When chronic tracheitis and bronchitis are associated with the laryngeal affection, inhalations and the topical application of menthol in olive oil, or creasote, guaiacol (2 per cent.), salol (20 per cent. to 30 per cent.) in olive oil by intra-laryngeal injections, have yielded excellent results in the hands of Grainger Stewart, Bronner, Sharp and Downie. About ʒj of the oily solution is injected daily straight into the trachea by means of a special syringe, preferably Bronner's silver and glass syringe. Care must be observed to pass the nozzle of the syringe well below the glottis during a deep inspiration, before injecting the contents.

In the treatment of *pachydermia laryngis* there is great diversity of opinion. If the use of alcohol or tobacco has been excessive, the habit must be corrected; if the laryngeal condition is associated with chronic pharyngeal catarrh or laryngitis, treatment appropriate to these conditions must be pursued. Rest and mild astringent laryngeal sprays will produce a favourable result in slight cases. Internally proto-iodide of mercury in gr.$_{\frac{1}{6}}$ doses, three times daily, has appeared to me to materially lessen the deposit, especially when combined with a laryngeal spray of tincture of thuja occidentalis (1 in 10). The administration of small doses of iodide of potassium has many advocates. Scheinmann has obtained good results from the use of steam inhalations of a 2 or 3 per cent. solution of acetic acid for ten minutes, three times daily, for some weeks. O. Chiari, to prevent recurrence, recommends electrolysis, as employed by Moll, viz., a current of ten to twelve milliampères for from three to five minutes at a time.

CHORDITIS TUBEROSA.

Chorditis Tuberosa, or "singer's nodule," is a clinical variety of pachydermia. A peculiar, small pale poppy-seed-like growth appears on the upper surface and free border of a vocal cord, about the junction of the anterior third with the posterior two-thirds, and is surrounded by a zone of hyperæmia. On the opposite cord there is often a corresponding depression. These nodules are the result of persistent over-use of the voice, and are most commonly seen in sopranos. When we remember that in the production of very high notes in the small register the posterior two-thirds of the cords are held in apposition, and only a small chink left anteriorly, we can see a reason for this particular nodal point being subject to constant attrition in sopranos. It is remarkable how comparatively little the voice may be affected; indeed, I once had a patient who, although he was a distinguished tenor, had continued to sing up to the time he came to me, in spite of the nodule and hyperæmia of the cords shown in the figure. After a period of absolute rest, and the constant application of thuja and other local treatment, he completely recovered, and only then realised how much his voice had been affected. His chief difficulty was in passing smoothly from the upper to the middle register, and the lower notes were impaired most.

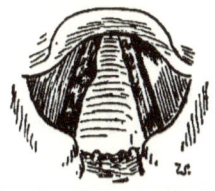

FIG. 27.
Chorditis Tuberosa.

Treatment.—This requires considerable patience and perseverance, if any permanent benefit is to result. The first essential is rest. When congestion has subsided the nodule may be removed by a sharp curette and a mild local astringent solution, *e.g.*, choride of zinc, 20 grs. to the oz., applied daily for a week. Mackenzie advised simply the application of perchloride of iron solution. Rice advises removal by a small snap-shot guillotine and chloride of zinc; while Bosworth prefers nitrate of silver (30—60 grs. to the fl. oz.), and avoids the use of forceps. I rely chiefly on the use of thuja, internally and locally, and the proto-iodide of mercury, as in chronic laryngitis, and the correction of any faulty method in singing.

Chapter XI.

MEMBRANOUS CROUP AND DIPHTHERIA.

MEMBRANOUS CROUP; DIPHTHERIA; INTUBATION.

MEMBRANOUS CROUP.

Etiology and Pathology.—While there is some reason to regard idiopathic membranous croup as a definite and distinct affection, it must be remembered that caustics or irritants, such as strong ammonia, chromic acid, scalds, the galvano-cautery, will cause a membranous exudation indistinguishable from the membrane of the idiopathic affection. A membranous laryngitis has also been observed in cases of scarlet fever, small pox, and typhoid fever, and whether the rare occurrence of a *membranous* exudation in these affections is simply due to the fact that the inflammation has occurred in a mucous membrane modified by the exanthem, or is due to some special toxine, is an open question, though recent research (Councilman, Klein, and others) seems to point to the streptococcus as the etiological factor in non-diphtheritic membranous exudations of the throat and nose. There is no longer any doubt as to the occurrence of membranous croup distinct from diphtheria.

Membranous croup usually attacks children between the ages of two and eight. It is most commonly caused by exposure to cold damp, and is more frequent in the country than in town.

Symptoms.—Membranous croup is generally primary in the larynx, but the false membrane may originate in the fauces or pharynx, spreading thence to the larynx. It may begin with the 'usual symptoms of a catarrhal laryngitis, but in the course of an hour or two, if not more suddenly, a characteristic loud brassy cough comes on, and gets worse especially towards midnight. The child becomes restless, and the temperature gradually rises till it reaches 102° or 103°. The cough, at first occasional only, becomes more frequent, and the attacks are often followed by laryngeal spasm. In the space of a few hours the symptoms of laryngeal obstruction appear, and the child endeavours to cough out the obstructing matter, clutches at its throat, and is more restless. Before long the cough is characteristically croupy, and the respiration becomes more and more embarrassed as the membrane is formed in the larynx, till, owing to its presence, the voice and cough become aphonic (silent croup), and the air forced through the narrowed glottis produces the peculiar crowing sound.

It is not unusual for some remission of the symptoms to occur towards morning, so that the acute stage may not be reached till the evening of the second or third day, but the course of membranous croup differs from the catarrhal form in its being *progressive*. When the acuter stage has been reached the cough may be frequent, and possibly may result in getting rid of some of the obstructing membrane, with temporary relief to the breathing. When the glottic obstruction is well marked there is great inspiratory and expiratory dyspnœa, and asphyxia may be threatened, necessitating intubation or tracheotomy.

At first the pulse is hard and frequent and the affection pursues a sthenic course; but if the obstructive dyspnœa continues, the deficient supply of oxygen and the carbonic acid toxæmia cause a fall of temperature and rapidly progressive weakness which makes the case very difficult to differentiate from diphtheria.

The Prognosis is exceedingly grave; it may terminate fatally in twenty-four hours, or may last seven or eight days. Hilton Fagge gave the mortality at 60 to 70 per cent., while Morell Mackenzie put the recoveries of cases untreated by tracheotomy as low as 10 per cent. Of 505 cases of true croup, collected by McNaughton and Maddren, and treated by calomel fumigations, intubation, or tracheotomy, 54·5 per cent, recovered; while of 275 cases treated by calomel fumigations alone, 48·7 per cent. recovered.

Diagnosis.—Membranous croup requires to be differentiated from spasmodic croup, laryngismus stridulus, diphtheria and retro-pharyngeal abscess. Regarding the differentiation between croup and diphtheria, we may lay special stress on the fact that " in the graver disease we have a thick, yellow, efflorescent false membrane, closely adherent to the parts beneath, and which cannot be separated from them without the rupture of blood-vessels, and furthermore the exudation at the end of twenty-four hours shows marked evidence of necrosis. The croupous membrane, on the other hand, is an entirely superficial deposit. The special diagnostic point lies in the fact that it is easily detached from the parts beneath without the rupture of blood-vessels " (Bosworth).

Treatment.—The difficulty of altogether satisfying oneself that any case is not really diphtheria renders it advisable to treat every case as though it was diphtheria, and all measures that tend to exhaust the patient should be studiously avoided.

At the outset it is well to administer a non-depressing emetic, such as ipecacuanha wine or powder. The patient should be placed in a steam bed, and hot fomentations applied to the throat, for such measures aid in the expulsion of the membrane. My usual practice is to administer calomel in doses of one to two grains, according to the age of the child, repeated every two or three hours till the bowels have acted freely, and then at less frequent intervals. The administration of calomel in membranous croup, advocated by Dundas Grant many years ago, is endorsed by Lennox Browne and by Bosworth, who favours equally the exhibition of hydrarg. c. cret. When the mercurial salt, having acted on the bowels, is being given at longer intervals, perchloride of iron should be prescribed. Pilocarpine cautiously administered is advocated for inducing diaphoresis, and loosening the membrane, but it is a dangerous remedy. Spraying the larynx can only be adopted in older children. Peroxide of hydrogen spray is praised by Glascow: "The liberation of gas which takes place seems to raise the membrane from its attachment, and so facilitates its expulsion." I have no personal experience of its use in membranous croup.

Dr. Corbin's or the Brooklyn method of treatment by calomel fumigations has yielded excellent results in the practice of many American physicians, and is advocated by O'Dwyer, Northrup, Maddren, and others. They recommend a dose of one or two grains of calomel before the sublimations, and the sublimation should be begun early, as soon as the diagnosis of true croup can be made. The patient is placed in a completely enclosed steam tent bed, and the calomel takes about ten minutes to volatilise, the tent bed being kept closed for fifteen minutes. "A very safe method is to volatilise grs. xv every two hours for two days and nights, then prolonging the intervals to three hours on the third day, four hours on the fourth day, fumigating three times daily thereafter according to indications" (O'Dwyer). If pure calomel is used, ptyalism rarely occurs, but "after prolonged use there is more or less anæmia. This must be controlled by administration of iron, and if there is associated prostration a little whiskey may be required before fumigation" (Northrup).

When the laryngeal obstruction becomes acute, the patient must be intubated or tracheotomised. The relative merits of these alternative procedures are discussed on pages 102 and 103.

DIPHTHERIA.

As diphtheria is treated at length in every text book of general medicine no extended notice of the disease is called for here, our remarks being chiefly confined to the laryngeal affection and its treatment.

Etiology and Pathology.—The essential factor in diphtheria is the Klebs-Löffler bacillus, and, unless the specific bacillus is present, the disease is not diphtheria. In the description of membranous croup I referred to the occasional formation of false membrane in scarlet fever, smallpox and typhoid fever, and as the local results of scalds and powerful irritants. The false membrane of croup, and that produced by scalds and caustics, is composed of coagulated fibrin, and disintegrating leucocytes and epithelium. In the false membrane of scarlet fever streptococci and a few staphylococci abound, but in diphtheria alone is the Klebs-Löffler bacillus found. Klein, as the result of his recent researches, confirming the conclusions of Babes, Holzinger, and others, lays down the following general rule:—
"Diphtheritic symptoms appearing as a sequel to scarlet fever, *i.e.*, in one to two or three weeks, are true diphtheria; but the like symptoms occurring as a complication, *i.e.*, in the course of the fever, are not, and though the streptococcus of scarlet fever may be present we shall not find 'Löffler's bacillus.'"

Detection of the Specific Bacillus.—The importance of examining the exudation for the presence of the specific bacillus has therefore an important bearing on the diagnosis. Occasionally the bacillus may be detected by direct microscopical examination of the exudate, but as a rule it is necessary to prepare a culture. Many methods and culture media have been advocated. Of these one (Sydney Martin's) may be mentioned. If any membrane can be seen, remove a small piece with sterilised forceps, and place it rapidly in a sterilised blood-serum cultivating test-tube. If no membrane can be obtained, a little of the mucus from the throat may be taken and streaked in a test-tube with a platinum wire, previously sterilised by heat. The tube is then put in an incubator and left at a temperature of 38° C. If the bacilli are present, a typical growth will be obtained on the surface of the test-tube medium within twenty-four hours. Other micro-organisms that might be present do not grow with sufficient rapidity to form a growth in the time, but a microscopical examination of the culture renders the diagnosis absolute.

Diphtheria varies greatly in intensity, and though Roux, Yersin, and E Fränkel believe that the pseudo-diphtheritic bacillus (Hoffmann Löffler's) is simply an attenuated form of the Klebs-Löffler bacillus, which under proper conditions may regain its virulence, Escherich and the German school generally regard the pseudo-bacillus as a distinct microbe.

The bacillus may be present without the formation of membrane. This, however, is a rare occurrence, and it is probable that a false membrane in these cases is often present in some parts inaccessible to observation.

The primary seat of the diphtheritic membrane is generally in the fauces. the velum, pillars of the fauces, or the tonsils; very rarely it occurs as a primary deposit in the nasal passages or the larynx.

Laryngeal diphtheria is very rarely primary. Thus Northrup reports that in 151 cases of diphtheria, in only one was the deposit confined to the larynx. The symptoms of the primary laryngeal diphtheria are at the outset almost indistinguishable from those of membranous croup; the temperature as a rule is lower, being only slightly above normal, while the early presence of albumen in the urine and marked constitutional depression should be regarded as conclusive evidence of diphtheria.

Since the laryngeal affection is almost invariably associated with and secondary to faucial diphtheria, there is seldom any room for doubt as to its real nature. The danger is that the absence of more acute signs of laryngeal obstruction, owing to the great depression of the vital powers, should lead one to over-look, or under-estimate, the evidence of extension of the membrane to the larynx. An examination of the larynx, if obtainable, would reveal the grayish false membrane lying on the slightly hyperæmic mucous membrane.

Nasal Diphtheria is attended with nasal obstruction, and a copious discharge of muco-pus, and occasionally streaks of blood. The discharge is very irritating to the skin of the nasal orifice and upper lip, producing redness and excoriations on which false membrane forms. The escape of fluid food by the nose, from imperfect action of the soft palate in faucial diphtheria, should not be taken as conclusive evidence of extension of the membrane to the nose. Owing to the rapid absorption of the virus by the vascular nasal mucosa, it is an extremely fatal complication.

Treatment.—The diphtheria bacillus does not enter the blood vessels, but acts by generating a chemical product, or toxine, which is absorbed and gives rise to constitutional symptoms. Treatment should therefore be directed (*a*,) to prevent the inoculation of other individuals by careful isolation, scrupulous cleanliness, and disinfection of all articles that may have been contaminated; (*b*,) to maintain the strength of the patient by avoiding all unnecessary exertion, by the frequent administration of light, easily assimilated food, and small quantities of alcoholic stimulants; (*c*,) to combat the general, and especially the cardiac depression, by the internal administration of such remedies as iron, quinine, strychnine, and strophanthus; (*d*,) to prevent the extension of the membrane and to destroy the vitality of the specific microbes, so as to prevent formation and absorption of the poison; (*e*,) to overcome the obstruction to respiration when the membrane has extended to the larynx.

For the general, internal, dietetic and hygienic treatment of diphtheria, the reader is referred to text-books of general medicine.

Of the recently introduced antitoxin treatment of diphtheria, Eastes, at the Meeting of the British Medical Association at Bristol, stated his belief that "in the early stages it will be found to be almost perfect." He had treated four cases himself with most encouraging results. Ehrlich, Kossel, and Wassermann, from 220 cases treated with a mortality of 44·9 per cent., urge the importance of early injection; cases treated on the first and second day of the disease yielded recovery in 100 and 97 per cent. respectively; cases treated on the fifth day, the recoveries were 56·5 per cent. only. The antitoxin is apparently harmless and doses of 5 c.c. to 20 c.c. should be injected, and, if no improvement has resulted, repeated on the same or following day according to the severity of the case.

As regards the *local treatment* of the parts affected, the general opinion is that it is useless to remove the false membrane, as it quickly re-forms in spite of every endeavour to prevent its doing so. The local treatment that in my experience has proved most satisfactory is frequent spraying with a 1 in 1,000 solution of perchloride of mercury, in water and glycerine, equal parts. The mouth and fauces should first be cleansed with a 3 per cent. solution of Condy's fluid, using a coarse spray. A fine spray should be used *per os*, as it insures the perchloride solution reaching every part of the fauces. Some prefer painting the parts thoroughly. The process of cleansing and spraying should be repeated every three hours, till the membrane disappears; then once or twice daily for some days, carefully noting the re-appearance of any fresh membrane as an indication for more frequent use of the germicide. It is necessary to watch for any indications of mercurial poisoning—a rare event. Insufflations of iodoform, or flowers of sulphur, have also been most serviceable. Pagan Lowe advocates a spray of equal parts of spir. vini rect. and water.

In nasal diphtheria a warm solution of salt in water (ʒj to the pint), followed by perchloride of mercury solution (1 in 3000—4000) should be used three or four times daily with the patient lying on one side. The nozzle of the douche is gently placed in the nostril and the stream allowed to flow gently in, returning by the other nostril.

F. H. Williams, as the result of an elaborate investigation and extensive experience, advocates 50 volume solutions of peroxide of hydrogen as a local germicide, either as a spray, or swabbed over the affected parts. Unfortunately, the strong solutions, from their acidity, cause some pain, lasting one minute. We may disguise the acid by adding sugar. The advantages claimed are: (1,) the strong solution is a good germicide, yet non-poisonous nor harmful to the mucous membrane; it cleanses a foul throat, and disintegrates certain portions of the membrane; (2,) it assists in diagnosis, for even a weak solution causes any trace of membrane to assume a white colour. Speaking generally, Williams advises that the solution containing $\frac{1}{2}$ per cent. of the acid should be applied every four hours. The 25 volume solution may be used as a spray; the 50 volume solutions may be applied a drop or two at a time on a swab till the membrane is removed or much diminished. It is well to use cocaine before applying the peroxide.

Inasmuch as great difficulty arises, from tonsillar crypts and cavities, in thoroughly disinfecting the parts, J. Macintyre has for some years past, after explanation to the patient, excised the tonsils in acute infectious diseases of the fauces, with the view of getting the germicidal agents properly applied. Where efficient nursing has been at his disposal to see that the raw surfaces were efficiently attended to, he has had most encouraging results even in diphtheria. Papain and zymine have been advocated as solvents. The combination of papain and lactic acid (Formula 48) is especially useful.

Lactic acid solution (50 per cent. and upwards) may be carefully applied to the membrane, acting both as a solvent and disinfectant. Liquor ferri perchloridi, iodine (40 per cent.), chloral, and sulphurous acid, are suitable applications. The galvano-cautery, and all powerful caustics and strong carbolic acid, should be avoided. Calomel fumigations, as in membranous croup (see p. 93), are advocated by American physicians in laryngeal diphtheria.

Acute Laryngeal Obstruction.—When false membrane has formed in the larynx and calomel fumigations or other means have failed to relieve laryngeal obstruction, the question of tracheotomy or intubation demands immediate consideration. Opinion is divided as to the advisability of early operation. Laryngeal diphtheria is extremely fatal; firstly, because such an extensive deposit is sure to involve more rapid absorption of the toxines, especially as the deposit is more or less beyond the reach of effective local germicides; secondly, because the membrane rarely stops at the larynx, but extends down the trachea and bronchi; thirdly, because it may cause death by closure of the glottis. Therefore, unless the dyspnœa is really marked, the operation is very often altogether useless, and one hesitates to advise an operation which cannot prevent the death of a patient from heart failure, or from blocking of the bronchi.

On the other hand, the performance of tracheotomy greatly facilitates effective local treatment of the infra-glottic extension of false membrane, and we may therefore be enabled to prevent further extension, and, in a few cases, pull the patient through. If the patient is not extremely weak, I should advise early tracheotomy in a

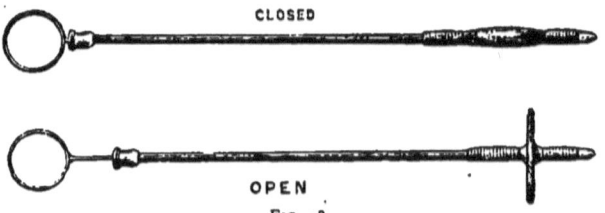

Fig. 28.
Tracheal Probang as used at the Bristol Royal Infirmary.

patient above five years of age. Speaking generally, under all other circumstances, I should prefer intubation as soon as dyspnœa supervened. If tracheotomy has been done, the local treatment should be carried out through the tracheotomy wound. If the membrane is extending down the trachea, it should be removed as far as possible by forceps, or by the tracheal probang, which some of my surgical colleagues have found of the greatest service, and small quantities of the perchloride solution may be injected through the tracheotomy wound.

INTUBATION OF THE LARYNX.

O'Dwyer's operation is in most cases the best method for the relief of laryngeal stenosis. His tubes are made of gilt metal, and vary in length from one and a half to

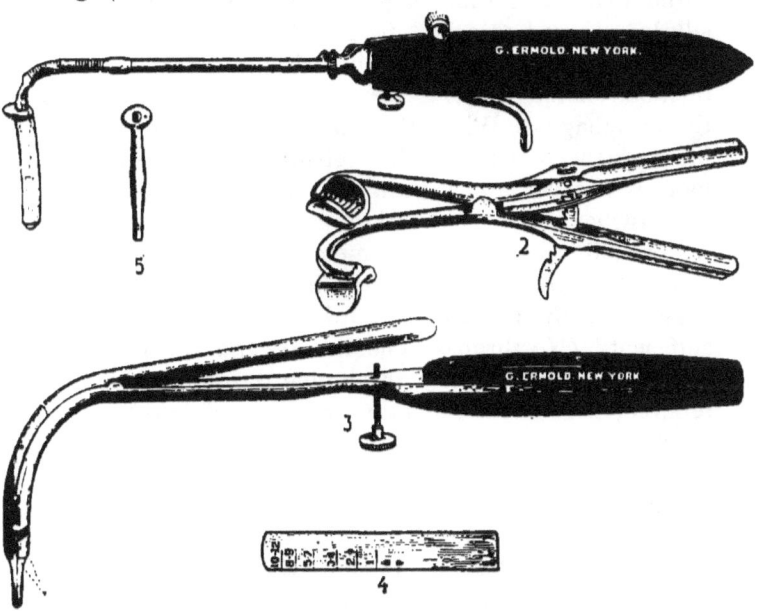

FIG. 29.
O'Dwyer's instruments for intubation.

two and a half inches for children of different ages. A flange at the upper extremity of the tube rests on the glottic margin with the tube *in situ*, the rest of the tube being below the vocal cords.

For intubation, the little patient having been firmly wrapped in a light blanket, the arms being included, is held sitting upright on the nurse's lap, and the mouth held open with the gag by an assistant.

The operator, facing the patient straight, hooks forward the epiglottis with the *left* forefinger. The selected tube is then quickly introduced on the obturator held in the *right* hand, and a sliding rod on the obturator handle then pushes the tube off the obturator, which is at once withdrawn. The tube is now in position, but with a silk loop-thread attached to the flange (previous to introducing), by which it may be withdrawn for re-introduction, should it have been inadvertently passed into the œsophagus. When the tube has been successfully inserted, the patient coughs violently a few times. The loop is not cut and withdrawn "until it is evident that all obstruction to the breathing has been overcome and no partially detached false membrane is in the trachea below the tube. The thread at first acts as an inciter to cough, which is desired; ordinarily ten minutes is sufficient" (Northrup). The larynx and trachea soon tolerate the foreign body, the urgent dyspnœa is at once relieved, the patient generally dropping off into a calm sleep.

Extraction of the tube is more difficult. For this, specially constructed forceps with closed blades are guided by the *left* forefinger tip into the upper orifice of the tube, opened so as to hold the tube firmly, and then withdrawn. Neither introduction nor extraction of the tube should occupy more than fifteen seconds, respiration being of necessity temporarily suspended during their performance. The position of the child is the same as for introduction.

The advantages of the operation over tracheotomy are:—

(a,) Its simplicity and painless nature, well illustrated by the fact that a child, seven years old, who had on former occasions experienced intubation and extraction at my hands, sat up and permitted me to extract the tube without being held, or in any way restrained. On this account we can resort to intubation much earlier than the more formidable operation of tracheotomy; and avoid all "cutting," to which the parents sometimes will not consent.

(b,) In children under five years of age, the percentage of recoveries in published cases is considerably higher than after tracheotomy, while above this age the percentages of recovery are about the same.

(c,) The intubation tube is worn much more comfortably than a tracheotomy tube; in fact, when *in situ* it can seldom be felt by the patient at all.

(d,) Coughing is more effectual; expectoration, therefore, more perfectly performed.

(e,) The respired air passes through the natural passages.

(f,) No anæsthetic is required. Cocaine is generally very useful.

Accidents and difficulties that may occur with intubation are:—

(1,) Pushing of false membrane into the trachea on introducing the tube. This occurred only three times in O'Dwyer's first 209 cases. As the thread is left attached till this danger is passed, it is a simple matter to remove the tube and allow the loosened membrane to be expectorated.

(2,) Coughing out the tube, with immediate return of dyspnœa—a rare occurrence.

(3,) Asphyxia, from blocking of the tube by false membrane. This can only occur in an extremely feeble patient, as the tube is always expelled by a vigorous cough, when, as occasionally happens, it gets blocked up. I have known the tube coughed up and swallowed in two cases without ill effects.

Northrup states that the symptoms of loose membrane are : (*a*,) croupy character of cough, the tube being in ; (*b*,) flapping sound ; (*c*,) sudden obstruction to out-going air, especially during coughing. When this condition is suspected, the thread should not be cut, but looped over the ear, protecting it along the cheek with adhesive plaister. Short tubes (for loose membrane) have been devised for these cases. They are short hollow cylinders of large calibre, and being short will not push down the tracheal membrane. As they have no retention swell, it is necessary to use the largest possible size.

(4,) Asphyxia from œdema above the tube and closing the orifice is a possible, but improbable, occurrence, inasmuch as the flange rests on the ary-epiglottic folds.

(5,) Ulceration at the cricoid ring from an ill-fitting tube.

(6,) The inspiration of particles of food and consequent pneumonia. This is the only untoward accident that I have experienced in twelve cases, and it can be obviated by using nutrient enemata while the tube is in.

(7,) Careless and forcible introduction may produce a false passage.

(8,) The special danger attending extubation is the tearing of tissues from opening the forceps widely at the side of, instead of within, the tube.

(9,) Asphyxia from prolonged, unskilled attempts at introduction or extraction is avoided by making several attempts, if necessary, rather than one prolonged one.

(10,) Difficulty may arise from sub-glottic stenosis at the cricoid ring, the narrowest part of the respiratory passages. The tube may meet with obstruction here, and even creep back, "like an oiled cork in a bottle." A smaller tube should then be used.

Intubation should be performed early so as to prevent engorgement of the lungs, and their partial collapse consequent on prolonged dyspnœa. Children who can understand should have nourishment in the form of thickened fluids, which they should "gulp" slowly. Casselberry's method of making the child swallow with the head low down and on its face has the effect of making the child gulp in this way. For those who cannot do this, nutrient enemata alone are preferable. The tube may be left in for five days or more, if necessary. It should be replaced as often as necessary. I have generally found it desirable to remove and clean the tube at least once every other day, if this can be accomplished without disturbing the patient to any great extent. Temporary aphonia sometimes persists for a few weeks after the removal of the tube.

As regards recent statistics of intubation and tracheotomy, it will be enough to quote the results of the collective investigation of the Deutsche Gesellschaft für Kinderheilkunde*:—

In four years 1,324 cases of primary diphtheria with laryngeal stenosis have been treated by intubation, with 516 cures (39 per cent.). The method is only applied in cases of severe stenosis. It is applied to all degrees of age, and in the youngest children. Of secondary diphtheria, following measles and scarlet fever, pneumonia, etc., 121 cases have been treated, with 27 cures (22 per cent.). In the first year of life 93 cases have been treated, with 13 cures (14 per cent.); in the second year of life 295 cases, with 92 cures (32 per cent.) Of 1,324 cases of in-

---

* Ranke *Münchener Med. Woch.*, 1893, No. 44, *Journ. of Laryng.*, Feb., 1894.

tubation, in 242 cases it was necessary to perform secondary tracheotomy, with 20 cures (7 per cent.); 58 cases with secondary diphtheria and secondary tracheotomy all died.

Of 1,118 cases treated by tracheotomy, 435 have been cured (39 per cent.). Of 42 cases of secondary diphtheria, 11 have been cured (26 per cent.). In primary diphtheria both methods gave exactly the same results, *i.e.*, 39 per cent. cures. In the first two years of life the results of intubation are better than those of tracheotomy.

I would add that it is impossible to draw comparisons between the relative advantages of intubation and tracheotomy from a simple comparison of results as exemplified by statistics, for whereas tracheotomy is seldom resorted to except in extreme cases, intubation is, or ought to be, undertaken as soon as laryngeal obstruction has manifested itself. It is also impossible to conceive that the large array of cases tabulated by several observers in any way represents the class of case that would have been deemed suitable for tracheotomy. We must rely rather on the general impression, as to the relative merit of these alternative procedures in those who have had wide experience in both of them.

Chapter XII.

THROAT AFFECTIONS OF INFECTIOUS FEVERS, GOUT, AND RHEUMATISM.

ENTERIC FEVER, SCARLATINA, MEASLES, SMALL-POX, CHICKEN-POX, INFLUENZA, GOUT, RHEUMATISM.

ENTERIC FEVER.

ERYTHEMA of the pharynx and fauces is not uncommon in the earlier stages of enteric fever, but though by no means a constant lesion, its occurrence is of no prognostic import. Occasionally the inflammation of the pharynx is attended with an herpetic eruption on the mucous membrane; this is a somewhat painful condition, but it usually subsides spontaneously. Enlargement of the adenoid tissue in the pharynx, and of the tonsils, is stated by Bartholow to take place at the same time as the thickening and deposition in the Peyer's patches in the intestines, and occasionally white elevated patches are seen on the tonsils or posterior wall of the pharynx, which go on to ulceration or necrosis. Osler records a fatal case in which there was irregular ulceration in the posterior wall of the pharynx, 2 cm. by 5 cm. leading directly into the submucous tissue, an aggravated form of the pharyngeal ulcers described by Murchison as having a round, oval, or irregular outline, from two lines to three-quarters of an inch in diameter. Croupous pharyngitis may occur apart from true diphtheria, and is generally of the worst augury. Thus Morell Mackenzie, in his work, refers to six cases reported by Oulmont, five of which terminated in death, whilst Peter states that all the instances he has met with have proved fatal.

Laryngitis in enteric fever occurs in two forms, the acute and the chronic.

Acute Laryngeal Complications in enteric fever are of more frequent occurrence than is generally supposed. No doubt in many cases the semi-comatose state of the patient explains the absence of subjective symptoms which would attract attention to the larynx. I have observed one case, referred to below, in which enteric fever commenced with the laryngeal affection, and two similar cases are recorded by Schuster. As a rule, the laryngeal affection does not develop until the fifteenth or sixteenth day.

Wilks has pointed out that there is a great tendency for these inflammatory changes in the larynx to undergo ulceration, and Hoffmann observed ulceration in twenty-eight cases out of two hundred and fifty, while in two thousand autopsies at Munich they were noted in one hundred and seven cases (Osler).

As to the nature of these ulcerations, there is difference of opinion. Fagge, Murchison, and most authorities regard them as secondary lesions, and not in any sense primary or specific lesions of enteric fever. On the other hand, Mackenzie, Rokitansky, and others contend that they are specific typhoid ulcerations, and I am strongly inclined to take the latter view ; and as this is a point of considerable clinical importance, I will briefly allude to three cases of enteric fever coming under my observation, which appear to support such an opinion.

Ernest S., age 20, was admitted to the Royal Infirmary on the sixth day of his illness. He became very delirious, and on or about the seventeenth day there was well-marked laryngitis and considerable bronchial catarrh, and he was continuously coughing and expectorating about the bed. He improved somewhat, but had a relapse, and on the twenty-ninth day of his illness he developed symptoms of acute laryngitis with dyspnœa. A steam bed relieved him, but in the evening he died suddenly, probably from the rapid occurrence of œdema. At the autopsy, in addition to the usual *post-mortem* appearances of *enterica*, the epiglottis and arytenoid folds were œdematous, while the ventricular bands were extensively ulcerated, the ulcers being purulent and sloughing. The right vocal cord had an ulcer on the *processus vocalis*.

Fourteen days after the death of this patient one of the nurses who had been attending him developed symptoms of enteric fever. She had not been in contact with any other case of enteric fever, and every precaution in dealing with the fæcal evacuations had been rigidly observed. The disease ran a peculiarly virulent and fatal course.

But another patient in a different ward, the only one who was under my care, began to develop febrile symptoms about the same time as the nurse. He was a man, aged 38, who had been in the ward for some months for aortic aneurism. The earliest symptoms of his attack of enteric fever were intense cephalalgia, and, a few days later, laryngitis and bronchitis. I found some hyperæmia and superficial ulceration of the vocal cords about the eighth day, but he had had syphilitic laryngitis years before, and his larynx had been permanently damaged. He developed typical symptoms of enteric fever, but on the ninth day he became very cyanosed and collapsed from respiratory failure, and not due to laryngeal obstruction, and died in a few hours. At the autopsy, in addition to all the typical lesions of enteric fever in the second week, there was superficial ulceration on the *processus vocalis* of each cord and on the anterior surface of the arytenoid cartilages.

Now the question arises as to the manner in which this patient contracted the disease, for he was in a separate ward, and one in which there was no other patient with enteric fever. He was rarely, if ever, visited by any friend from without, but I found that he had been in the habit of going to visit a patient in the ward where the delirious patient, Ernest S., was, although he never went within a few feet of E. S. I cannot avoid the conclusion that both my patient and his nurse contracted the disease from the expectoration of Ernest S., who was proved to have had typical typhoid lesions of the larynx while he was expectorating about the bed clothes.

Additional evidence of the correctness of this hypothesis was furnished by cultures of the Eberth-Gaffky bacillus in agar tubes inoculated from the ulcers on the arytenoids, and from the spleen.

These cases would, therefore, seem to explain the possible infectiousness of typhoid fever, as maintained by Budd, a view endorsed by Collie,* and brings home to us the necessity for more careful prophylaxis in cases exhibiting laryngeal complications. They also support the views of Landgraf and others who contend that the laryngeal ulcers of typhoid fever may be true typhoid ulcers.

The occurrence of ulcerations has also been attributed to mechanical causes acting on the inflamed and infiltrated mucosa. Dittrich, more especially, considers that they are the result of mechanical pressure of the larynx against the posterior pharyngeal wall, or arise from attrition.

---

* On Fevers, p. 53.

Œdema of the larynx is a rare complication. Osler in recording one fatal case refers to the exhaustive article by Lüning,* wherein he states that œdema was present in nine out of a hundred and fifteen autopsies of enteric fever cases in which there were serious laryngeal complications.

Perichondritis with subsequent exfoliation of the cartilages is liable to follow ulceration, and the arytenoid cartilages or the cricoid may be denuded and become necrosed from this cause. A thin pseudo-diphtheritic pellicle sometimes forms on the epiglottis and within the larynx.

The *chronic form*, according to Peter, does not commence until convalescence has been established, generally about two months after the attack of enteric fever has terminated. The voice becomes hoarse, there is a difficulty in phonation, and œdema is liable to supervene. This is an exceedingly grave complication, and very liable to terminate fatally, but even if the patient survives after tracheotomy, the condition is a very troublesome one to relieve, as cicatrices form and contract, producing a very persistent stenosis of the larynx necessitating retention of the cannula.

Prognosis.—The milder complication of catarrhal angina and laryngitis does not appear to modify the course of the disease any more than does a catarrhal bronchitis, but ulceration of the larynx is a very grave complication, the percentage of fatal cases being very high, while very few indeed of the diphtheritic cases survive. The prognosis of the chronic form of laryngeal ulceration is scarcely less grave, especially if œdema occurs. According to Sextier, of ten cases which necessitated tracheotomy, not one survived.

---

* Die Laryngo-und Tracheostenosen im Verlaufe des Abdominal typhus. *Arch. für. klin. Chir.* Band xxx.

Treatment.—In the simple catarrhal class of cases, soothing inhalations or the sucking of ice, and drinking cool fluids will afford relief, while the fauces and pharynx may be sprayed with a solution of permanganate of potash. When ulceration has occurred the only treatment that is likely to succeed is counter-irritation by sinapisms or blisters externally. For œdema, tracheotomy will have to be performed, but, if the patient survives, the resulting cicatricial stenosis will generally necessitate the tube being permanently retained.

The virulent type of enteric fever with which these lesions are associated generally precludes any active therapeutic measures directed to the larynx being successful.

SCARLATINA.

It is scarcely necessary to dwell at any length on the pharyngeal affections in scarlatina, since they constitute such a very essential feature in most cases of this fever, and are fully described in text books of general medicine.

As the result of an elaborate investigation conducted by Walter Dowson, of Bristol, on the *rôle* of the tonsil in scarlatinal infection, he is able to bring very strong evidence tending to prove that we should regard the tonsillar lesion characteristic of scarlet fever as the cause rather than the merely symptomatic consequence of a specific septicæmia, of the existence even of which we have no satisfactory evidence; that, in fact, as in diphtheria, the throat affection and cervical bubo of scarlet fever are the analogues of the chancre and inguinal bubo of syphilis, or of the intestinal ulcers and enlarged mesenteric glands of enteric fever.

Scarlatina anginosa is the variety in which the tonsillar and faucial inflammation is a marked feature.

" The throat affection may be serious from the first; but more frequently, in a case which presents no very unusual features at the beginning, it undergoes aggravation either at the acme of the fever, or during the subsidence of the rash, or even its disappearance . . . There may be abscess of the tonsil, or ulceration and gangrene, with œdema of the surrounding tissues ; and supervening thereon, the glands in the neck may inflame and suppurate, and sinuses form" (Bristowe). With deep ulceration fatal hæmorrhage may occur. The inflammation in these anginose cases may extend to the larynx and the resulting infiltration and œdema produce aphonia, while painful deglutition occurs if the epiglottis is involved. The larynx, however, is seldom much affected, and œdema and ulceration here is much less frequent than in either measles or enteric fever.

A croupous laryngitis may arise by extension from the fauces—a very dangerous complication. Though diphtheria often follows scarlatina, Klein has found that a false membrane occurring in the course of the fever is not truly diphtheritic.

Later, and occurring as a complication of general anasarca, non-inflammatory œdema of the larynx may occur, and is liable to prove very rapidly fatal.

There is a great tendency for the pharyngeal lesions to spread by the Eustachian tube to the middle ear, but this complication is by no means limited to cases of scarlatina, and is very liable to supervene in measles and other exanthemata with marked pharyngitis.

Treatment.—In the less severe cases, simple soothing and antiseptic gargles and sprays are useful, while in the more aggravated throat affections, poultices or hot fomentations should be applied externally and the fauces swabbed out or sprayed with some powerful antiseptic, such as biniodide of mercury (1 in 2000), listerine, or sanitas (1 in 6). A tonic treatment and highly nourishing diet are called for in the anginose form of scarlatina.

The possible necessity for tracheotomy should always be borne in mind.

MEASLES.

(1,) A certain amount of catarrhal pharyngitis and laryngitis is present in almost every case of measles, the severity of the throat complications varying greatly but being influenced considerably by family predisposition, and tending to be especially severe in some epidemics. The swelling of the laryngeal mucosa may produce great dyspnœa, as well as the usual hoarseness and cough.

(2,) In the stage of invasion, the so-called "endanthem," a mottled, punctate rash of the soft palate, fauces, tonsils, and, according to Löri, of the laryngeal mucous membrane, is often a valuable diagnostic sign before the appearance of the characteristic cutaneous rash.

(3,) Spasmodic laryngitis or false croup is not unusual in young children in the early stages.

(4,) Membranous laryngitis, apart from true diphtheria, may occur in measles, generally coming on late in the attack as the cutaneous rash declines. It is a very dangerous complication, of course, the prognosis and symptoms being similar to the idiopathic form. Dr. Sam. West observes that "this variety of croup seldom begins until the eruption of measles is on the decline, or the process of desquamation has commenced.

(5,) Laryngeal ulcers are occasionally met with on the vocal cords, ventricular bands, or posterior laryngeal wall (Gerhardt).

(6,) Gangrenous laryngitis is rare, and is generally

associated with gangrene of the mouth, in ill-nourished, unhealthy children.

(7,) H. Smith records three cases in which paralysis of the intrinsic muscles of the larynx occurred as a sequel to a mild form of measles. The paralysis set in a few days after the subsidence of the fever, and lasted from six to ten days. In each case laryngoscopic examination revealed total bilateral paralysis of the abductors and adductors of the vocal cords.

Treatment.—For the simple catarrhal and membranous laryngitis the treatment is the same as for the idiopathic forms (see p. 77 and p. 92). For gangrenous cases no local treatment appears to be of any use.

SMALL-POX.

The pharynx and larynx are inflamed to a slight extent in most cases. Rühle in fifty-four autopsies found the larynx affected in all. In many cases the vesicles appear on the fauces, and these have been observed in the larynx as discrete, well defined white spots on the epiglottis, arytenoids, and vocal cords, and in the trachea and bronchi. The symptoms of pustules in the larynx are those of laryngitis, and generally come on about the sixth day, and a few days later œdema may supervene, but Rühle observed that the essential feature of the laryngeal complication was a diphtheritic membrane far more often than pustules. The formation of a false membrane generally commences about the tenth day. Interstitial laryngeal hæmorrhage often complicates the hæmorrhagic variety of small pox, while perichondritis and muscular paralyses are also liable to arise.

Treatment.—The treatment of these complications of small-pox are the same as for similar idiopathic affections of the larynx. Cicatricial stenosis is liable to follow extensive ulcerations occurring during the course of the fever.

CHICKEN-POX.

The characteristic vesicles surrounded by a well marked areola may sometimes occur on fauces, tonsils, etc. After a few days, superficial ulcerations form and heal spontaneously.

INFLUENZA.

An acute catarrh of the pharynx is an essential feature in most cases of influenza, and the larynx and bronchi are generally affected to a less extent. In a few cases the inflammatory infiltration is œdematous, or breaks down, or is attended with the formation of a false membrane. Shelley has described a vesicular eruption on the soft palate, the vesicles appearing like small well boiled sago-grains. The tonsils may be much inflamed and swollen. A notable feature of all the throat complications of influenza is their tendency to be followed by a persistent chronic congestion, and Cohen considers that there is undoubtedly a vaso-motor paresis of both the blood vessels and lymph vessels.

Œdema of the larynx has been observed by Wolfenden and Pugin Thornton, and Cohen states that hæmorrhage and abscess of the larynx are common. As in the pharynx, the acute trouble is very liable to be followed by chronic laryngitis. By extension from the naso-pharynx the ear is often implicated in the acute catarrhal inflammation. Instances of acute inflammation and of empyema of the antrum of Highmore, due to influenza, have been noted.

Treatment.—The treatment of the throat affections of influenza does not differ from similar inflammatory conditions of the pharynx and larynx from other causes.

GOUTY AFFECTIONS OF THE THROAT.

The throat manifestations of gout may assume the acute or chronic form, and both gout and rheumatism are undoubtedly a very frequent cause of throat affections.

The lithæmic diathesis has been referred to elsewhere as an etiological factor in acute and chronic pharyngitis. There is usually a well-defined, bright scarlet, patchy hyperæmia of the mucous membrane, which is dry.

In gouty *angina tonsillaris* the patient generally complains much of the pain. The tonsil does not suppurate, but the intense redness and soreness may yield suddenly to an acute articular attack. Similarly, *painful laryngitis* is suggestive of gout. I have seen a case in which nocturnal laryngeal spasms occurred whenever an error in diet rendered the patient gouty.

FIG. 30.
Gouty Pharyngitis.

Duckworth, in his treatise on gout, states that "the gouty throat is like no other. The pillars of the fauces, especially the posterior pair, the velum, and the uvula, are very red and glazed. They appear as if freshly brushed with glycerine. Some dilated venules may often be seen coursing over parts of the membrane. . . . The surface of the pharynx is coarse, with red, glairy prominences upon it, and depressions, here and there, covered with grayish, slightly adherent patches of mucus, and it has sometimes enlarged venules upon it." He refers to the observation of de Mussy, who had a case of granular pharyngitis in which masses of concretion, consisting of carbonate and urate of lime, were discharged several times daily from the mucous follicles; while Virchow has detected a small gouty concretion in the posterior part of a vocal cord.

Harrison Allen, Hinkel, and others describe also a patchy congestion of the laryngeal face of the epiglottis, extending along the aryteno-epiglottidean folds and over the posterior aspect of the ventricular bands, together with a harsh, dry cough, with a sense of extreme irritation in the larynx. Hinkel states that when the patchy inflammation is present there is extreme sensitiveness to astringent and stimulant applications, which in itself is a point of diagnostic significance. This patchy condition may exist in the pharynx, extending as streaks along the postero-lateral walls, with a sense of uneasiness or pain on swallowing. The pain darts into one or both ears, and, as he puts it, "seems to come out of the ear—to be a very long pain, in fact, apparently extending beyond the surface of the body."

Morell Mackenzie reported four typical cases of gout in the throat, in all of which there were other proofs of their true gouty nature. These cases were (1,) an acute œdema of the uvula, disappearing upon sudden development of an ordinary attack of podagra; (2,) a chronic inflammation of the posterior pillars of the fauces; (3,) a gouty deposit around the crico-arytenoid joints on both sides, causing permanent dysphonia; (4,) a gouty inflammation producing fungous ulcerations of the left ventricular band, resembling cancer so strongly, objectively as well as in the subjective symptoms, that both Krishaber and M. Mackenzie had suspected it to be cancer, and discussed the possible necessity for extirpation of the larynx. This candid statement by such an experienced observer shows most clearly how great the difficulties may be in arriving at a correct diagnosis.

A form of *rhinitis sicca* is sometimes observed in gouty patients.

Treatment of these gouty manifestations is that of gout generally, the only local treatment necessary being some sedative spray or lozenge, such as Formulæ No. 40 and No. 8.

The following is a favourite prescription with me:—

℞ Tinct. colchici sem - - - - - ♏xx
   Sodii salicylat. - - - - - grs. xx
   Tinct. digit. - - - - - - ♏iij
   Aq. dest. ad - - - - - - ℥ss

To be taken in a tumblerful of Vichy water (Celestin), twice or three times daily.

Of course, suitable dietetic rules must be laid down and rigidly adhered to.

RHEUMATIC AFFECTIONS OF THE THROAT.

That acute tonsillitis and acute pharyngitis is very frequently a manifestation of rheumatism is now universally admitted and widely recognised by medical practitioners. But cases of acute laryngeal rheumatism are not very rare. There is nothing peculiar in the appearance of rheumatic inflammations; the chief point is to bear in mind the possibility of both acute and chronic pharyngitis and laryngitis being rheumatic, since successful treatment will depend entirely on a correct diagnosis of these cases. Arch. E. Garrod[*] states that "it is true of the rheumatic sore throat, as of many other rheumatic manifestations, that there are no distinctive

---

[*] A Treatise on Rheumatism.

features by which its nature can be recognised in any given case." He quotes Fernet to the effect that, " If the throat be examined, a diffuse erythematous redness is seen to occupy the whole of the back of the throat, and some œdematous swelling of the mucous membrane is present, which is most marked about the uvula, which is swollen and elongated. The pharynx is moist and free from all exudation ; one or both tonsils may present more or less considerable swelling."

I have met with cases of acute laryngitis which have immediately preceded, and yielded to, an attack of acute tonsillitis.

Chapter XIII.

CHRONIC INFECTIVE DISEASES.

SYPHILIS; TUBERCULOSIS; LEPROSY.

SYPHILIS OF THE LARYNX.

Inherited Syphilis of the larynx occurs: (1,) within the first few months of life, when it usually takes the form of laryngeal catarrh, or the milder manifestations of secondary acquired syphilis; (2,) about puberty; in this later form, tertiary manifestations may be encountered.

Acquired Syphilis of the larynx assumes the character described as secondary and tertiary (one case of primary chancre is reported by Moure), but "secondary" manifestations may arise and keep on recurring many years after the primary sore; and on the other hand "tertiary" forms may be met with rarely within a few months of the initial lesion. While this statement applies especially to laryngeal syphilis, it is likewise true of faucial and nasal syphilis.

Secondary manifestations take the form of:—

(1,) Simple catarrh, and erythema.

(2,) Mucous patches, condylomata, and superficial ulceration.

In tertiary syphilis of the larynx we may find:—

(1,) Diffuse infiltration and superficial ulceration.

(2,) Gummata and deep ulceration.

(3,) Epithelial outgrowths.

(4,) Perichondritis.

(5,) Cicatrices and fibroid contractions.

Syphilitic Catarrh in no way differs from simple non-syphilitic catarrh, except in its persistency. Whistler and others refer to the patchy appearance of the hyperæmia. The history and the occurrence of syphilitic lesions in other parts generally enable one to make a correct diagnosis.

Mucous Patches and Condylomata are not often seen. In fact early syphilitic disease of the larynx gives rise to such slight symptoms that it is frequently unsuspected, and for this reason the tertiary form of the disease, with more pronounced symptoms, more often comes under the notice of the physician.

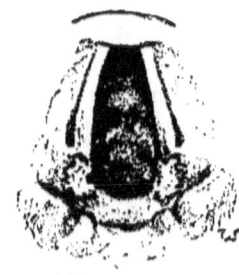

FIG. 31.
Bilaterally symmetrical superficial ulcers on the vocal processes in secondary syphilis.

The circumscribed gray thickening of the infiltrated epithelium may occur on the epiglottis (especially on the lingual surface), the ary-epiglottic folds, posterior commissure, or on the vocal cords. They are generally single, or if multiple are not symmetrical. Superficial erosions, yellow, oval, circumscribed, and surrounded by an areola, may follow denudation of the softened epithelium, especially in professional voice users. The symptoms are hoarseness, and very slight expectoration.

Diffuse Infiltration leading to tumefaction of the epiglottis, inter-arytenoid fold, or vocal cords may cause considerable hoarseness, the "raucous" voice, and sometimes dyspnœa. The infiltration may break down, forming chronic superficial ulcers, or may undergo fibroid transformation.

Gummata are sometimes seen before breaking down as smooth, red or yellowish, defined swellings; generally single and occupying the epiglottis (its margin or the laryngeal surface), the ary-epiglottic folds, posterior wall of the larynx, or the ventricular bands; or they may be infraglottic, or tracheal. When about to break down the centre becomes yellow and ulcerates. The whole gumma then rapidly disintegrates from the centre towards the periphery, and a characteristic syphilitic ulcer results. (See p. 67.)

Perichondritis generally occurs in association with gummata, either by deep extension of the infiltration or, more rarely, the infiltration may be seated primarily between the perichondrium and the cartilage.

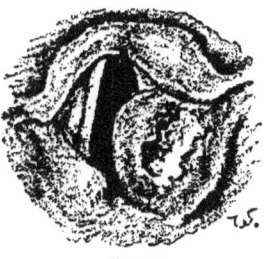

FIG. 32.
Breaking down gumma of the left ary-epiglottic fold.

In the latter case especially, necrosis, and destruction, or exfoliation of the cartilage is liable to follow.

In 1880 Semon drew attention to a sclerosing form of perichondritis in which a fibroid change occurred without any breaking down or caries of the cartilage. This form is chronic and persistent, and leads to marked stenosis and deformity.

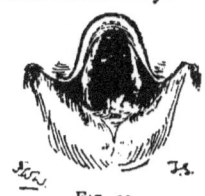

FIG. 33.
Syphilitic out-growths in the inter-arytenoid space.

Neoplasms or mammillated outgrowths are often found projecting from the posterior commissure. They resemble those found in tubercular laryngeal disease, but consist of proliferated epithelium. They are sometimes associated with ulcerations.

Fibroid Contraction of the larynx follows diffuse infiltration, which undergoes the fibroid metamorphosis, and gives rise to marked stenosis. Cicatricial contraction is as characteristic of syphilitic ulceration of the larynx as of the fauces. A web may be formed between the cords, as in the case illustrated.

Fixation of the Cords from ankylosis of the arytenoid cartilages is a frequent consequence of fibroid thicken-

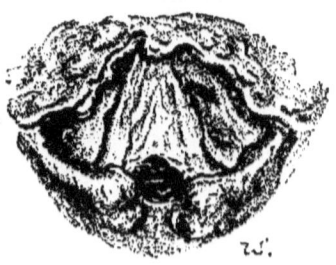

Fig. 34.

Extensive destruction of the epiglottis, and formation of cicatricial web between the vocal cords. The result of tertiary syphilis.

ing in this region. *True paralysis* of the vocal cords from nuclear disease or from pressure on the motor fibres of the nerves to the larynx in the brain or in the nerves at the base of the brain, may occur, and is sometimes the earliest physical sign of a diffuse syphilitic lesion of the brain. (See p. 163 *et seq.*)

Symptoms.—The chief symptom in laryngeal syphilis is hoarseness. While the absence of pain is remarkably characteristic of syphilitic lesions, here, as in the fauces, it is not safe to rely on this feature too absolutely. A gumma of the epiglottis may cause considerable dysphagia, or if on the posterior surface of the cricoid cartilage, much pain may be felt on deglutition. The peculiar raucous voice of advanced syphilitic laryngeal disease often enables one to make a diagnosis before examining the larynx. The voice is generally affected, to what extent depends, of course, on the extent and seat of the disease. There is often great difficulty in distinguishing the cord at all, in cases where the ventricular bands have been much altered and infiltrated, and with the vocal cords ulcerated and more or less completely gone, or cicatrised and adherent to the ventricular bands.

Treatment.—The general treatment of syphilitic lesions in the throat is practically the same as for syphilis occurring in other regions, and therefore need not be discussed at length. It is necessary, however, to emphasise the importance of not adhering rigidly to routine methods for the so-called secondary and tertiary forms of the disease, for we shall often find it necessary to administer iodide of potassium in the secondary form, and mercury in the later manifestations.

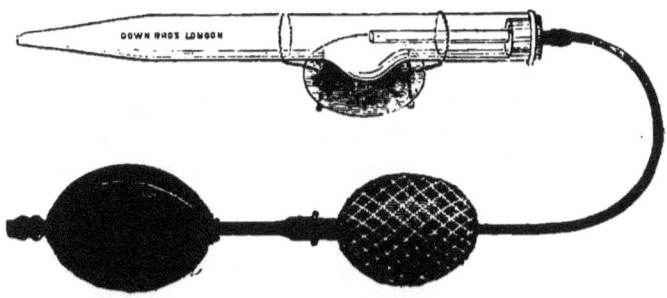

FIG. 35.
Calomel Fumigator for laryngeal and nasal syphilitic ulcers and for calomel fumigation in membranous croup.

Mucous plaques, however, often give much trouble before they disappear. As a rule, the application of solid nitrate of silver, or of the saturated solution, carefully limited to the affected areas, should be made every alternate day until the patches begin to resolve—combined, of course, with internal antisyphilitic remedies. Where there are cracks or erosions, or when pain is present, Browne has usefully substituted iodine and carbolic acid as a local pigment. In obstinate cases McBride advises painting with chromic acid (grs. x. ad ʒj).

The foul ulcers of tertiary syphilis may require a gargle of chlorate of potash, or some other mild antiseptic; or the application of a solution of sulphate of copper or carbolic acid. Iodide of potassium should be given internally in daily doses of 60 grains from the commencement; in the later manifestations, gummatous infiltration and ulceration and perichondritis should be rapidly brought under control by free administration of the iodides, combined with mercury. If ulceration is extensive and progressive, the calomel fumigator may be used with advantage.

Stenosis of the larynx, if chronic, may necessitate tracheotomy or intubation. As a rule tracheotomy is to be preferred, as syphilitic stenosis is so liable to recur after dilatation by Schrötter's bougies or intubation tubes. If the stenosis be due to fixation of the vocal cords, it may be possible to mechanically dilate the glottic opening, and to keep them diverged by means of Störk's dilator till they have become fixed in the new position; or one cord may be excised.

Fig. 35.
Störk's tracheal cannula and dilator for chronic laryngeal stenosis.

Felix Semon in reference to the treatment of chronic fibroid stenosis, strongly advocates thyrotomy with subsequent excision of the cicatricial tissue, especially in syphilitic cases in which the voice is already lost and tracheotomy has been performed.

O'Dwyer has urged the advantages of intubation, and several of his cases have yielded most brilliant results. Often only small tubes can be passed at first, but after leaving these in for twelve or twenty-four hours it is generally possible to introduce larger ones, and eventually obtain a permanent stretching of cicatricial tissue.

Cicatricial web formations should be divided by the cutting dilator, and, as advocated by Max Thorner, intubation tubes worn till the edges have healed, so as to obviate reunion and re-formation of the web.

TUBERCULAR LARYNGITIS.

The records of Brompton Hospital show that tubercular disease of the larynx is found in at least 50 per cent. of all patients dying of chronic pulmonary tuberculosis; whilst about 20 per cent. of patients suffering from phthisis manifest signs of tubercular laryngeal disease (P. Kidd).

Etiology and Pathology.—Primary laryngeal tuberculosis may undoubtedly occur, but the affection is usually secondary to tuberculosis of the lung, although physical examination may not reveal the lung disease, for the physical signs, in any but advanced cases, tend to be obscured by the throat affection. It occurs more frequently in men than in women, in the proportion of about 2 to 1.

The tubercular affection probably originates by direct infection by the sputum passing over it. A long standing or frequently recurrent catarrhal laryngitis often precedes the true tubercular deposit, and possibly predisposes to infection by weakening the resisting power of the tissues, or, by causing small superficial erosions, affords a means of entry for the bacilli.

We may distinguish three stages in the development of the tubercular lesions:—

(a,) The *first stage* is accompanied by pricking sensations in the throat, irritability of the fauces and larynx, and general langour. The mucous membrane of the

larynx and pharynx is usually remarkably pale even at this stage before any local deposits of tubercle can be observed. But pallor is not always present; in fact, the larynx may be simply congested.

(b,) In the course of a week or two the *second stage of deposition* manifests itself. Generally, pale

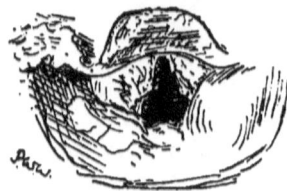

Fig. 37.

Tuberculosis of the larynx, with ulceration of the epiglottis and infiltration of the ary-epiglottic folds and ventricular bands.

Fig. 38.

Turban-shaped infiltrated epiglottis and pear-shaped arytenoids from tubercular infiltration.

smooth swellings of the epiglottis, or arytenoid regions appear, at first unilaterally, soon becoming bilateral, till the turban-shaped pale grayish epiglottis and the pear-shaped swelling of the arytenoids

Fig. 39.

The same as Fig. 38, but two months later, showing more extensive infiltration and rapidly advancing ulceration of the tubercular deposit.

present the characteristic aspect of tuberculosis of the larynx. *(See Plate III., Fig. 2.)* In some cases miliary tubercles may be observed forming in the pink swollen mucous membrane, as grayish yellow points which are obviously lying beneath the translucent mucous membrane, and which rapidly increase in size and number till by coalescence they form the typical pale gray infiltrations more commonly seen.

(c,) The *third stage of ulceration* rapidly succeeds the tubercular infiltration. Tubercular ulcers are superficial, often multiple, with irregular "mouse-nibbled" edges, difficult to define from the surrounding pale gray infiltration, and covered with pale grayish-white *débris*. They tend to spread slowly and superficially rather than deeply. There is no attempt at cicatrisation of tubercular disease.

Again, indurative tubercular laryngitis, with slight tumefaction and little ulceration, may persist for many months, and somewhat suddenly take on the more typical and rapidly progressive characters of laryngeal tuberculosis. This indurative form often presents the mammillated proliferations already alluded to. (*See Plate III., Fig. 4*).

Paresis or paralysis of the vocal cords is not uncommonly due to the associated laryngitis inducing paresis of the thyro-arytenoidei muscles, or to waxy degeneration of the muscles. True paralysis of a vocal cord may be the result of pressure on the recurrent laryngeal nerves; thus the nerve may be implicated in pleural thickening at the apex of the lung, generally the right, or an enlarged gland may press on either nerve.

While the epiglottis and the arytenoid regions are the favourite seats of tubercular disease, the ventricular bands or inter-arytenoid folds may be the first part of the larynx to be affected, the other regions remaining unaffected for a considerable time. Eventually the perichondrium of the cartilages may become involved with resulting perichondritis and caries or necrosis.

The vocal cords may be attacked first, becoming red

and swollen, the tumefaction breaking down and forming ulcers on the cords, without any other part of the larynx being affected. Ulcers on the vocal cords are common in advanced tubercular disease of the larynx.

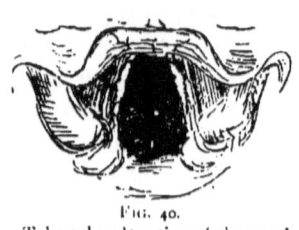

Fig. 40.
Tubercular ulceration of the vocal cords.

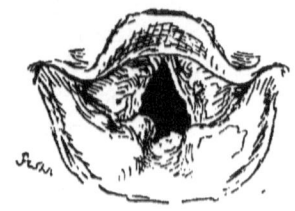

Fig. 41.
Mammillated tubercular outgrowths in the inter-arytenoid space, with infiltration and ulceration in other parts.

Fungous, villous, or mammillated outgrowths are sometimes observed, sprouting frequently from the inter-arytenoid fold, or from the vocal cord, or rarely in any other region; the outgrowths sometimes cover the whole arytenoid fold. It should be remembered that very similar outgrowths of a non-specific nature may occur in long-standing simple laryngitis, and in syphilis, but Störk and Mandl consider that their presence in the inter-arytenoid region is almost pathognomonic of tubercle.

The Symptoms of laryngeal tuberculosis will obviously vary greatly according to the character and extent of the lesions. Only slight hoarseness may be noticed in early cases, but when more advanced the most prominent features are cough, pain on deglutition associated with rapid loss of flesh, a hectic temperature, and great prostration.

The constant worrying cough is due to the extremely irritable condition of the larynx, the amount of expectoration depending mainly on the extent to which the lungs are involved.

Pain on swallowing is present when the epiglottis is infiltrated, and is often agonising when it has undergone ulceration. Perichondritis, of the cricoid especially, is attended with pain on swallowing. The swelling of the epiglottis or arytenoid regions may prevent closure of the glottis, so that fluids find their way into the larynx and bring on attacks of painful coughing. As a consequence of the intense pain and difficulty attending swallowing, patients will often refuse almost all food, while the saliva and secretions are allowed to accumulate and dribble from the mouth.

The degree of hoarseness depends on the implication of the vocal cords, while inter-arytenoid swelling prevents their approximation, and results in aphonia; or the arytenoid cartilages may be fixed by the tumefaction, with consequent hoarseness or aphonia.

The acuteness of onset and development varies greatly in different cases, and at different periods in the same case, but the affection of the larynx is always an extremely grave complication of tuberculosis. The onset in some cases is very insidious, and the infiltration and ulceration may occur without any symptom but hoarseness being complained of; but when we remember that 50 per cent. of patients dying from phthisis show laryngeal complications, the importance of attending to any laryngeal symptoms in pulmonary consumption is obvious.

Treatment.—The general treatment of laryngeal tuberculosis of course includes the general treatment of pulmonary tuberculosis, but one or two points require special notice. Firstly, the patient should be strictly enjoined to use the voice as little as possible; he may speak in a low whisper instead of phonating. All irritants, such as tobacco, strong alcoholic drinks, highly seasoned food, dusty occupations, etc., should be avoided:

food should be cool, bland and soft, and alcoholic stimulants well diluted.

As regards the question of residence, the general consensus of opinion is opposed to sending patients with laryngeal affections to high altitudes. There may be exceptional cases, of course, but we have the high authority of Theodore Williams for regarding laryngeal tuberculosis as a contra-indication for the rarefied dry air of the higher Alps. In fact, patients who are too far advanced for the active treatment of the laryngeal complication are, as a rule, better kept at home; or, if they are able to settle abroad, the northern African sea-board, or the Riviera in Europe, will afford the most suitable climatic conditions. On the other hand, if active therapeutical measures are to be carried out, it is better to postpone the question of residence abroad till the laryngeal disease has been arrested.

In discussing *local treatment*, we may divide cases into three classes:—

(*a*,) Those in which no definite tubercular deposit is observed, but only a catarrhal laryngitis, localised or general. For such, local measures consist in subduing the pain and cough, if present, by menthol or cocaine, and morphine or codeine in small doses, either applied locally or by a spray or pastil. Thus :—

℞ Menthol - - gr. xx to xl
   Cocainæ vel - -
   Morphinæ - gr. v
   Ol. vaselini - ℥ j

To be used in the oil atomiser before meals when the cough is troublesome.

℞ Menthol - - gr. ¼
   Codeinæ - - gr. ¼ vel
   Morphinæ - gr. 1/12
   Acid. citric. - gr. ½
   Saccharin elix. - q. s.
   Fiat pastil.

To be dissolved slowly in the mouth; three or four may be used daily if required.

A pastil, or the use of the spray at night, will often ensure a good night to the patient.

(*b*,) Those in which a definite tubercular deposit has occurred in the larynx.

If only a local tumefaction without ulceration is present, I have obtained excellent results from the injection into the affected mucous membrane of a few minims of a solution of perchloride of mercury in water and glycerine, 1 in 1000; or of pyoktanin, 2 per cent., held in solution; or of aristol, 2 per cent., in almond oil. A similar plan of treatment has been employed by Krause and others.

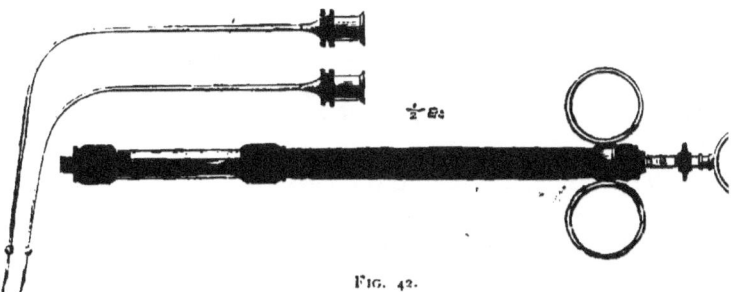

Fig. 42.

The author's Syringe for submucous injections in tuberculosis of the larynx (half size).

As the result of these injections, the pain is almost always diminished in the course of about twenty-four hours, and in several cases the mischief was arrested after repeated injections. Krause removes such tumefactions with his double curettes, and applies lactic acid; Gougenheim employs a very similar instrument, especially for arytenoidectomy, in cases attended with dysphagia.

If ulceration has occurred, lactic acid should be well rubbed into the ulcerated surface. I generally follow

Heryng's practice of first curretting the tubercular necrotic tissue before rubbing in the lactic acid. Krause removes the tubercular deposit by means of his double

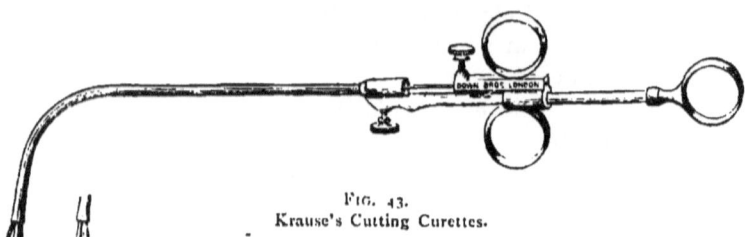

FIG. 43.
Krause's Cutting Curettes.

sharp spoon instruments, but, as I have remarked, does not confine himself to the removal of the ulcerated deposits.

The lactic acid solutions should be strong (50 to 80 per cent.), and must be applied by practised hands to the affected parts of the larynx only. In some cases such strong solutions of this powerful irritant are not well borne; the strength should then be reduced to 15 or 20 per cent.

Of course, all these procedures require a previous application of a 10 or 20 per cent. solution of cocaine, till the larynx is completely anæsthetised.

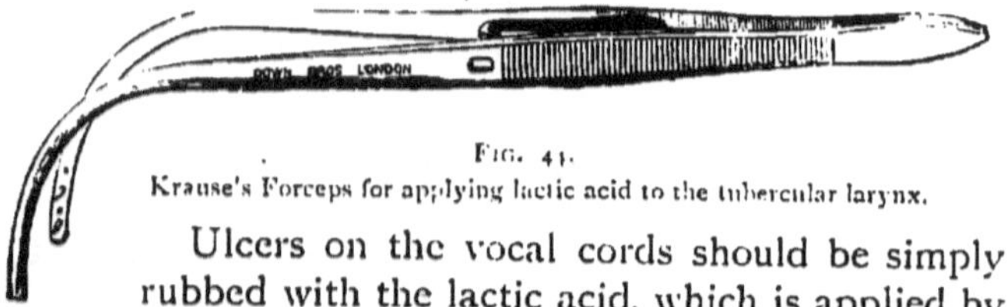

FIG. 44.
Krause's Forceps for applying lactic acid to the tubercular larynx.

Ulcers on the vocal cords should be simply rubbed with the lactic acid, which is applied by means of a small pledget of cotton wool firmly wound on a laryngeal probe. The patient should be confined to a warm room, and be directed to suck ice after the operations, and only to take cold and perfectly bland liquid food. Pain should be relieved by morphine, either hypodermically or combined with cocaine hydrochlorate in minute doses in the form of a pastil.

If any active inflammation or perichondritis is present, *curettement* and the application of lactic acid are generally contra-indicated, for the disease must be quiescent before such radical procedures can be undertaken with safety or any prospect of success.

In favourable cases the tubercular ulcers may be made to cicatrise by these means, and cases are recorded, and have occurred in my own practice, in which the larynx has remained free from tubercular disease for years.

To Heryng, of Warsaw, is mainly due the credit of devising and establishing the radical treatment of laryngeal

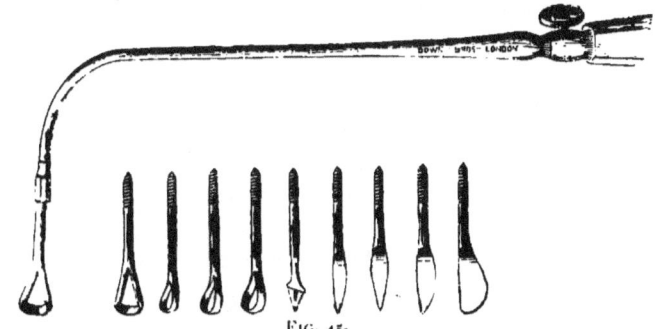

FIG. 45.
Heryng's Laryngeal Curettes.

tuberculosis by *curettement*, and in reporting his results based on 252 cases that he has himself treated he states that* *curettement is especially indicated :* (1,) In cases of circumscribed, slowly developing tubercular infiltrations, even though they may show no tendency to break down. The object of the operation here is to prevent the further inevitable destruction of a vital organ by destroying the centre of infection. Even somewhat advanced lung

---

* *Journ. of Lar. and Rhin.*, Aug. and Sept. 1893, May 1894.

disease and a certain degree of fever, so long as it is not of a hectic character, cannot in all cases be regarded as contra-indications. If the tubercular infiltration is confined to the posterior wall of the larynx, as is most frequently the case, then an early and as radical as possible removal of this can bring the process to a standstill for months and years, and restore the functions of the larynx ; (2,) In many diffuse infiltrations which run their course with special violence, and even when the general condition is relatively unfavourable—*e.g.*, excessive dysphagia due to inflammatory swelling and ulceration of the epiglottis or posterior wall of the larynx—for though, of course, healing cannot take place, the pain can be alleviated in the quickest manner possible for a considerable time.

The chief contra-indications Heryng considers to be (*a*,) Advanced phthisis of the lungs with hectic and wasting ; (*b*,) Diffuse miliary tubercle of the larynx, or rather of the larynx and pharynx ; (*c*,) All cachetic conditions ; (*d*,) Severe stenosis of the larnyx caused by inflammatory swelling of the affected parts. In these cases tracheotomy must be performed as soon as possible ; (*e*,) Patients exhibiting fear and nervous excitability, mistrust of a physician, and who are always changing their doctor, *especially in those whose condition promises little hope of recovery.**

Further, in speaking of the dangers which may be associated with the treatment, Heryng refers to (1,) the question of hæmorrhage. In 270 cases treated he has only seen severe hæmorrhage twice, in each case after removing a tumour-like hard false cord. He therefore considers it advisable to destroy hard tumour-like tubercular infiltrations of the false cords by means of electrolysis or

---

* *Journ. of Laryng.*, Aug. 1894.

the galvano-cautery. The subsequent treatment of the wounded surface with lactic acid after *curettement*, depends on whether we have succeeded in removing all the diseased parts.

(2,) The other question demanding consideration is the possibility of an outbreak of general tuberculosis being hastened or excited by a local operation in the larynx. Hitherto he has not witnessed such an event, though Lermoyez and Sokolowski each report an instance, but, as Heryng remarks, it is rather a difficult matter to prove the connection.

Other methods often afford relief, *e.g.*, simple insufflation of iodoform (Schnitzler) with or without tannic acid and morphine; or applications of 20 per cent. menthol in olive oil (Rosenberg); or a solution of 12 or 15 per cent. of menthol with 2 to 4 per cent. of guaiacol dissolved in olive oil and injected through the larynx into the trachea by means of a special syringe, as advocated by Downie.

(3,) Those cases which are associated with acute inflammation or perichondritis, or in which the rapidly advancing lung affection has greatly weakened the patient, can only be treated by palliative remedies such as local applications of morphine, cocaine, or menthol.

Tracheotomy is rarely called for, and should only be performed when dyspnœa is urgent. It is generally to be preferred to intubation, since it is only resorted to in cases where an intubation tube would not be well tolerated.[*]

LEPROSY OF THE THROAT AND NOSE.

The throat and nose are frequently affected in patients suffering from the cutaneous form of this

---

[*] For Lupus of the Larynx, see Lupus of the Pharynx and Larynx, p. 59.

disease, but it has never been known to occur primarily in the respiratory tract.

It is unnecessary to enter into the vexed question of the etiology of this affection beyond remarking that the lesions are due to a microbe so closely resembling the bacillus of tubercle, that at one time it was even regarded as a modified form of tuberculosis.

It may assume the tuberculated form, or, very rarely, the anæsthetic variety. In either case the onset is extremely insidious, owing to the painless nature of the affection, and patients will sometimes declare that they have nothing the matter with the throat, when examination reveals that it must have been developing for a considerable period.

Tuberculated Leprosy of mucous membrane passes through three stages. In the first stage the uvula and soft palate, in which the alterations are usually first observed, become red and velvety in appearance, the neighbouring tissues becoming affected by continuity or by separate foci of disease, so that the nasal mucosa, the epiglottis and aryteno-epiglottidean folds become red, velvety, thickened, and hard, like the soft palate and uvula, and appear as though coated with varnish. At this stage epistaxis frequently occurs, and the patient may complain of shortness of breath, a sense of tickling and dryness in the pharynx, larynx and nose.

In course of time, the red, hard infiltration becomes soft, and the tissues somewhat œdematous, the redness soon giving place to pallor till the affected regions are uniformly pale and resembling the anæmia of tubercular deposit, and when the infiltration and cell-

Fig. 46.
Leprosy of the Larynx (Wagnier).

ular elements become absorbed, the tissues appear, as Mackenzie puts it, as though they were infiltrated with tallow.

In the second stage, the characteristic tubercles appear, first as small, whitish-yellow, elevated nodules, varying in size from a pin's head to a split pea, isolated, or in chains or groups. They may remain stationary for years. Lennox Browne, who examined the cases at the Leper Establishment at Robben Island, found that it was generally impossible to see anything below the epiglottis, owing to the enormous thickening from infiltration of that cartilage; but Ramon de la Sota, whose valuable article in Burnett's system is based on a very extensive experience, states that on the appearance of the nodules the uniformly swollen condition of the membrane disappears.

The third stage of ulceration is rarely reached, as the patient generally succumbs to the general affection ere the nodules break down. The ulcers are rounded, elevated above the surrounding mucous membrane, and are compared by De la Sota to syphilitic mucous patches. The cartilages of the larynx are sooner or later involved and become necrosed, and may be exfoliated.

In the nose, the mucous membrane of which is often involved by extension from the alæ nasi, the chief symptom is nasal obstruction, but when the nodules ulcerate, the stench of the sanious watery discharge is often intolerable.

FIG. 47.
Leprosy of the Larynx (G. Mackern).

The cartilaginous septum becomes perforated, and, with the alæ nasi, may be destroyed by extensive ulceration.

The earliest indication of the extension of leprosy to the upper respiratory passages is alteration in the voice, which at first nasal in quality, soon becomes shrill. Later in the disease, hoarseness or aphonia results from the implication of the vocal cords. Dyspnœa is rarely very marked, though in a few cases tracheotomy is required for the stenosis of the larynx resulting from œdema or from the nodular infiltration.

The sense of smell and taste is usually blunted or lost early, and the loss of sensation is peculiar and complete. De la Sota has found that there is often complete anæsthesia without analgesia.

The *anæsthetic form* of leprosy is stated by Hillis, of Demerara, never to occur until the cutaneous disease has existed for at least five years. The mucous membrane is pale, the velum palati paralytic, and the affected regions are anæsthetic.

Diagnosis.—Leprosy of the throat and nose requires to be differentiated from syphilis, tuberculosis, lupus, and cancer. There is rarely any difficulty in making the diagnosis, in that leprosy is never a primary affection of the upper air passages, but always secondary to, or at least concurrent with, cutaneous leprosy, though a leprous patient may be affected with cancer, lupus, or syphilis of these regions.

Treatment can only be palliative, and is only called for in the ulcerative stage when alkaline and antiseptic solutions may be useful, or for the relief of dyspnœa by tracheotomy. De la Sota has noticed improvement under the application of a 1 per cent. solution of resorcin, and of iodoform dissolved in ether, and from touching the diseased areas with a 10 per cent. solution of chloride of zinc.

Chapter XIV.
NEOPLASMS OF THE LARYNX.
BENIGN NEOPLASMS; MALIGNANT NEOPLASMS.

Under neoplasms in the larynx are included the benign and malignant new growths, but not those vegetative forms of laryngeal syphilis and tuberculosis, nor the epithelial formations termed pachydermia laryngis, nor of course, inflammatory deposits, although clinically they constitute laryngeal tumours.

BENIGN NEOPLASMS.

The only common varieties of benign neoplasms are the *papillomata*, which occur more frequently than all the other forms together, although out of 56,498 patients in Schrötter's clinic only 230 cases of laryngeal papilloma were observed, and the *fibromata* constituting about 10 per cent. of benign laryngeal growths.

Etiology.—The real causes of benign growths in the larynx are most indefinite. While Lennox Browne believes that syphilis is an important predisposing factor in their production, Morell Mackenzie thought that the syphilitic and tubercular diatheses had precisely the opposite effect, tending to immunity.

Excessive use of the voice, especially in the open air and in dusty atmospheres, is generally held to be an important factor in their production, yet papillomata and other benign growths have occurred congenitally and in deaf mutes.

Papilloma.—This, the commonest variety of laryngeal new growths, is met with at all ages, but especially in young adults. It is almost the only form seen in young children. Papillomata may be single, but are generally multiple, varying in size from a millet seed to a walnut, of a delicate pink granular surface, with an almost characteristic cauliflower-like uneven surface. They are usually sessile, firm, elastic, and do not readily bleed (see *Plate IV., Figs. 1, 2*). In histological structure according to Virchow, all the papillomata are warts, consisting of epithelial cells, with a small formation of connective tissue.

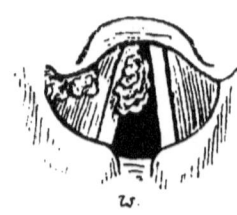

Fig. 48.
Multiple papillomata of the larynx.

The favourite site is the free border or upper surface of the middle or anterior half of the vocal cords (especially the right cord), or the anterior commissure. Unless multiple, they are less frequently seen on the ventricular bands, ary-epiglottic folds, or epiglottis. Unlike epitheliomata, their area is distinctly limited, they do not infiltrate the surrounding tissue, and are practically never seen in the posterior third of the vocal cord, or inter-arytenoid fold. Yet early epithelioma of the larynx may very closely resemble a benign papilloma (see *Plate IV., Fig. 3*).

Fibrous Polypus (the so-called fibroma) is a fairly common variety of tumour almost confined to adult life. Fibromata are generally single, sessile, with light pink or cherry-red, smooth surface, and are mostly found occupying the upper surface of the middle or anterior half of a vocal cord, and vary in size from a millet seed to a walnut. They may be pedunculated, and if large may present an indented uneven surface resembling a papilloma. O. Chiari, in his recent communication at the

PLATE IV.

Benign and Malignant Laryngeal Growths.

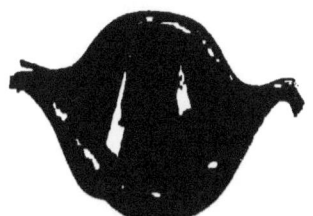

Fig. 1.

Fig. 2.

Fig. 3.

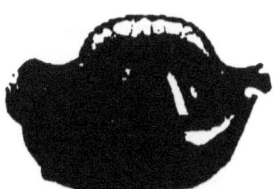

Fig. 4.

Fig. 1.—Papilloma laryngis in an adult female.

Fig. 2.—Benign growth attached to the under surface of the left vocal cord of an adult male. On attempted vocalisation the growth projected upwards, preventing closure of the vocal cords. Removal completely restored the voice.

Fig. 3.—Epithelioma of the larynx showing two separate warty growths. (For this drawing the author is indebted to Dr. P. M'Bride.)

Fig. 4.—Cancer of the Larynx.—An epithelioma is seen occupying the ary-epiglottic fold and extending to the arytenoid region and involving the ventricular band. It is bright pink and lobulated, but no portion had ulcerated when this drawing was made.

International Congress at Rome, has shown that the growths consist of the same tissue as the vocal cords, and originate in inflammatory thickening of the vocal cord from congestion, and thus they are vascular and contain cavernous blood spaces. Serous infiltrations and hæmorrhages are common, especially in the softer growths.

Cystoma results from obstruction in the duct of a muciparous gland, and consequently generally occurs where these are plentiful, especially the epiglottis. They are found also on the ventricular bands, or growing from the ventricle, the arytenoid region, and rarely from the vocal cords.

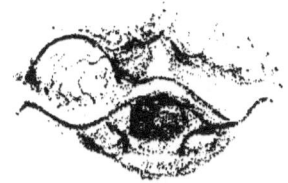

Fig. 49.
Cyst on the lingual surface of the epiglottis.

They are smooth, tense, globular, semi-translucent, and covered with light red or grayish-pink mucous membrane, and if considerable in size, blood vessels are seen coursing over the surface.

Angioma is a rare form of tumour, of characteristic aspect, generally unilateral and single, occurring in the sinus pyriformis, or on the ventricular bands, vocal cords or epiglottis, and rarely exceeding a filbert nut in size.

Myxoma usually occurs on the vocal cords as a small, smooth, pink or grayish-white, sessile, translucent tumour; unilateral and distinctly localised. If pedunculated it is rather a fibro-myxoma in structure, and then may present a mammillated surface and resemble a papilloma in appearance.

Ecchondroma is a very rare form of growth, arising generally from the cricoid cartilage. Ecchondromata have been found in connection with the epiglottis, thyroid and arytenoid cartilages.

They are firmly attached, hard sessile growths, presenting a smooth surface of irregular outline and covered with healthy mucous membrane.

Lipoma and adenoma are exceedingly rare forms of laryngeal neoplasm.

Prolapse of the Ventricle of Morgagni so closely resembles a neoplasm of the larynx in its clinical aspects that it is convenient to allude to it here. A smooth, pink, lobulated, supraglottic mass, either unilateral or bilateral, is seen resting on the vocal cords and corresponding to the opening of the sacculus, which of course is absent. It is most frequently associated with phthisis pulmonalis, and appears to result from atrophy of the thyro-arytenoidei muscles and to be brought about by violent coughing. As it is useless to replace it, the projecting portion should be snared or excised.

Symptoms.—Pain is hardly ever experienced, the symptoms being almost confined to impairment of voice, and a greater or less degree of obstructive dyspnœa, varying, of course, according to the situation and size of the growths.

In a certain number of cases no symptoms whatever occur, or they are so slight as to cause no discomfort to the patient. But even a small growth occupying the free border of a vocal cord, or the anterior commissure, may very greatly impair the voice or produce complete aphonia.

Cough is sometimes a marked symptom, particularly in young children with papillomata, in whom it may be croupy in character, especially as in them the growth is liable to excite some laryngitis and glottic spasm.

Treatment.—A course of full doses of arsenic combined with daily applications of tincture of thuja should first be undertaken in order to avoid if possible the necessity for operative interference. Papillomata are very liable to recur again and again if removed, especially in children, in whom, on the other hand, they sometimes disappear spontaneously as the child grows up. Even if skilfully and successfully removed, there is a considerable risk of some permanent impairment of the voice remaining. For these reasons Semon advises that in children these growths should not be operated on unless respiratory embarrassment is present. At the same time we must remember that papillomata in children may rather rapidly increase in size and number, and an operation become necessary.

In adults there is less liability to recurrence, and if the voice is impaired, or other symptoms are present, we should not hesitate to remove a papilloma. It has been maintained by some laryngologists that operative interference increases the liability to the occurrence of malignancy in benign growths, but any doubts on this point have been set aside by Semon's elaborate investigation, which proved that the liability of a benign growth to become malignant, while always excessively remote, is diminished rather than increased by skilful removal.

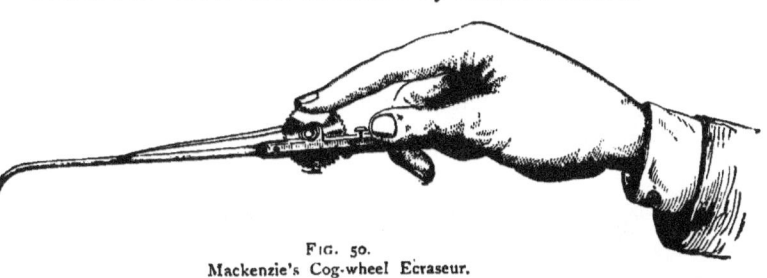

Fig. 50.
Mackenzie's Cog-wheel Écraseur.

As regards the choice of operation, the decision must depend on the nature and situation of the growth as to whether we use the sponge-probang, écraseur, curette, forceps, the cautery or knife, or resort to an external operation. Of course all intra-laryngeal operations require that the larynx and fauces should be well cocainised.

Small pedunculated growths on the free edge of the vocal cords may occasionally be removed by means of Voltolini's sponge probang, passed into the larynx and quickly withdrawn on closure of the glottis. I prefer using Heryng's curettes for such cases, as advocated by Massei, while Dundas Grant's guarded forceps are both

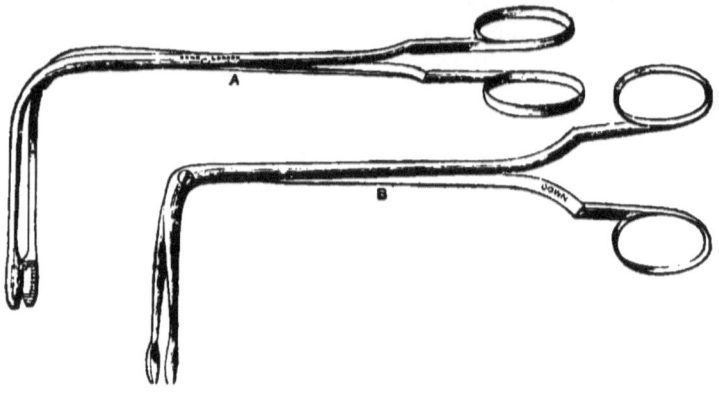

FIGS. 51 and 52.
Mackenzie's Cutting Forceps. A, the lateral; B, the antero-posterior forceps.

safe and suitable for growths on the free edge of the cords about the middle third, or in the anterior commissure. Mackenzie's cutting forceps is, in my opinion, the most generally serviceable instrument, while in

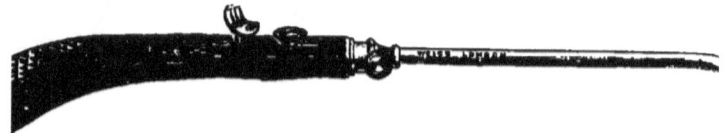

FIG. 53.
Schrötter's Forceps.

special cases Schrötter's, Wolfenden's or Gottstein's and other forms are to be preferred. For the larger growths snaring, when feasible, is preferable to picking away with forceps.

The galvano-cautery must be used with the greatest caution and skill, and is very rarely called for.

The removal of papillomata in children is said to be facilitated by Lichtwitz's fenestrated laryngeal tubes, the growth being made to project into the lumen of the tube through an opening corresponding with the position of the growth. Rosbach, of Wurzburg, has succeeded in removing supraglottic growths by means of a thin narrow knife made to penetrate the centre of the thyroid cartilage from without, on a level with the upper surface of the vocal cords, the intra-laryngeal movements being guided by laryngoscopy.

FIG. 54.
Gibbs' Laryngeal Snare.

After removal of a papilloma, it is well to paint the seat of the growth with tincture of thuja every few days, and to administer small doses of the proto-iodide or bin-iodide of mercury for some weeks. The excellent results obtained with thuja occidentalis in the treatment of warts by Kaposi and others would lead me to try its effect in cases of laryngeal papilloma unsuitable for operation, as well as with a view to prevent recurrence after removal. *Cystomata* should be seized with forceps. They always collapse, but the cyst-wall being partly removed, they do not recur. *Angiomata* may be cauterised with a galvano-cautery point at a dull red heat.

Of external methods, there are three alternatives :—
(*a*,) thyrotomy,* (*b*,) subhyoid pharyngotomy, (*c*,) laryn-

---

* *Syn.* Thyro-chondrotomy.

gotomy. It is claimed that the more complete removal of papillomata by means of thyrotomy lessens the liability to recurrence. Even if that be the case, the operation is not wholly devoid of danger to life, while the voice is almost invariably either lost or permanently impaired. It should, therefore, only be resorted to where asphyxia is threatened and the growths cannot be removed by the intra-laryngeal methods. For subglottic growths involving respiratory embarrassment, thyrotomy or laryngotomy is sometimes unavoidable, but there are very few cases of laryngeal benign growths requiring removal which cannot be reached *per vias naturales.*

MALIGNANT NEOPLASMS.

Malignant disease of the larynx is a fairly common affection from the age of 35 upwards.

Symptoms. — *Epithelioma* is encountered with far greater frequency than any other form of malignant disease. The primary growth usually commences on the vocal cords, the ventricular bands, or the epiglottis, but it is not confined to any one region, and in a large proportion of cases it is impossible to determine the primary seat of the new growth, as it rapidly infiltrates the tissues of the larynx.

The statistics of different observers as to the relative frequency of its appearance on the vocal cords, ventricular bands and epiglottis, vary too greatly to enable definite statements to be made on the point. On the vocal cords, in its earliest stages, epithelioma appears either (1,) as a pinkish warty growth, generally single, and bearing a strong resemblance to a benign papilloma ; or (2,) as a diffuse infiltrating growth, with a red uneven surface ; or (3,) giving an uneven fringe-like margin to the affected vocal cord. On the ventricular bands or ary-epiglottic folds it appears as a deep pink infiltrating growth, with a coarsely mammillated or uneven surface. The epiglottic growth is more grayish or whitish-pink, and looks almost fibrous in texture, but with uneven surface.

The growth may increase in size very slowly at first, but it soon extends rapidly and tends to ulcerate fairly early, or at least to become excoriated on the surface, and then readily bleeds. But deep ulceration is rarely long delayed. The floor of the ulcer is then covered with foul grayish mucopus and *débris*, tinged with blood, giving a characteristic musty odour to the breath.

As the growth and ulceration extend, a secondary perichondritis often complicates the disease and renders the laryngoscopic appearance less characteristic.

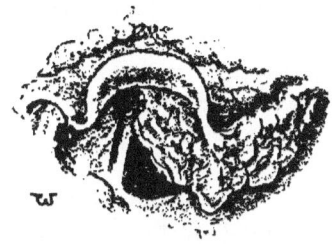

FIG. 55.
Epithelioma of the epiglottis. Laryngeal symptoms had been noticed for twelve months.

Krishaber, Butlin, and others classify primary malignant growths of the larynx into the *extrinsic*, affecting the epiglottis, arytenoids, ary-epiglottic folds and pyriform sinuses, and the *intrinsic*, including those arising in the vocal cords and ventricular bands, and the subglottic growths.

Presumably from the arrangement of the lymphatics, extrinsic growths tend to earlier secondary infiltration of the lymphatic glands of the neck than is the case with intrinsic growths. Mackenzie went so far as to say that the absence of secondary glandular infiltration was of little or no diagnostic value, and this is probably true early in the disease. Lennox Browne points out that in supraglottic deposits the lateral glands between the larynx and carotids and the subhyoid are often affected, and that in infraglottic cancer, while the glands of the neck escape, the pre-laryngeal and the lateral glands are often enlarged.

The larynx may be secondarily involved by extension of a malignant growth in the pharynx, but is scarcely ever the seat of secondary deposit from cancer in other regions.

Encephaloid Cancer is decidedly rare, and is difficult to distinguish from epithelioma of the larynx, except by microscopic examination. It is sometimes more vascular in appearance, it tends to involve the neighbouring lymphatic glands earlier, and to ulcerate more rapidly and more deeply than epithelioma, and occurs mostly in the epiglottis. I have never met with a case in my practice.

Sarcoma.—Round or spindle-celled sarcoma occurs in the larynx, but this form of growth is rare. It may commence in the epiglottis or ventricular bands, or as an undefined, diffuse, infiltrating growth. The growth is generally single, and, if defined, is smooth, globular and semi-translucent, but it frequently takes the form of a a grayish-pink, infiltrating growth, with smooth but uneven surface.

The rapidity with which it extends varies greatly in different cases, but it tends to remain unilateral, and in many cases the vocal cords escape, or are only implicated later in the disease.

The only case I have met with was a round-celled sarcoma, spreading from the thyroid gland to the larynx. It was observed as a subglottic infiltrating growth for three months before the voice was affected, and even then the growth seemed for some time to be retarded in extension upwards by the corresponding vocal cord.

Symptoms.—The symptoms vary, of course, with the size and seat of the deposit, and although there are certain features which point to malignancy, it is often extremely difficult to form a definite diagnosis in cases suspected to be malignant.

The *voice is almost always affected quite early* in the disease, either from inflammatory exudation or from infiltration of the cord, or of the tissues around the arytenoid cartilages, or from implication of the recurrent nerves by enlarged lymphatic glands, but on the other hand the voice is seldom altogether lost.

Pain is usually a prominent symptom, especially in extrinsic deposits, and is often severe, darting up to the ear, and increased by swallowing.

Cough is not as a rule a marked symptom. Loss of appetite and a general cancerous cachexia often supervene early. Respiratory obstruction depends on the size and situation of the growth.

Fauvel states that increased salivation (from reflex irritation) is generally present, and in consequence of the odynphagia, the saliva collects and is allowed to dribble out of the mouth.

Differential Diagnosis.—In the early stages a warty growth on the vocal cords or ventricular bands resembles a benign papilloma, but appearing late in life a unilateral growth would excite suspicion, especially if firmly fixed and infiltrating the subjacent structures, with marked impairment of vocal cord movement, and attended with pain. Its occurrence on the posterior third of the vocal cords, or on the posterior commissure, would point to malignancy.

A growth on the epiglottis or ventricular bands might be mistaken for a gumma. A gumma is very rapid in appearing, ulcerates very soon, and is rarely painful, and generally yields to iodide of potassium. When ulceration has occurred, especially with perichondritis, it is most difficult to decide from the laryngoscopic appearance between a tertiary syphilitic lesion and a breaking down malignant growth. The symptoms and physical signs of tubercular disease and lupus are usually sufficiently definite to prevent their confusion with malignant disease in the larynx.

In doubtful cases a piece of the growth ought to be removed and submitted to the microscope. This should not be a very minute piece, as the evidence afforded is often equivocal, and, in that case, the patient would have been better left alone. Newman goes so far as to say that no portion should be removed for diagnostic purposes unless the patient has consented to submit to a radical operation, if the diagnosis of malignancy is rendered certain on microscopical examination.

The difficulties in arriving at a definite conclusion as to the true nature of a growth, even after a microscopical examination, is well exemplified by a case reported by Sokolowski. He removed a polyp from the left vocal cord, believing it was benign. Four years later the growth having recurred, it was considered malignant and cauterised. Again, eight months afterwards, another laryngologist operated on it, and from the microscopical examination and the clinical history concluded it was a benign adenoma. But Sokolowski then examined the growth he had removed five years before, and concluded that it was an adeno-carcinoma, and this subsequently proved to be correct, as the growth recurred in a manner which left no doubt as to its malignant character.

Unless the diagnosis is unequivocal, the patient should always be put through an anti-syphilitic course, but it must be remembered that malignant growths are often rendered less painful for a time by iodide of potassium.

Treatment.—It is impossible to lay down definite rules for the treatment of these cases, opinions being so divided as to the relative advantages of various operative procedures that have recently been introduced, in some cases with most brilliant results.

The choice lies between palliative treatment and radical operation. In the latter case we have four alternatives: (1,) extirpation through the natural passages; (2,) thyrotomy and erasion; (3,) partial laryngectomy; and (4,) total laryngectomy.

Radical Operations.—A few cases of successful extirpation through the mouth are recorded; but only exceptionally, in early cases of intrinsic cancer, could this be attempted with any prospect of success.

Thyrotomy, with removal of the growth by excision and scraping, has been done in a good many cases with partial success, when the growth was unilateral. Partial laryngectomy probably affords the best chance of eradicating the disease, but can only be successful when the growth is strictly limited to one side, and the glands are not secondarily involved.

Total laryngectomy is the only operation when the disease has involved both sides of the larynx, but the operation is attended with many special dangers, and very rarely is the patient's life prolonged.

The methods of performing these operations and their relative merits are beyond the scope of this work, but it is well to bear in mind that in all there are added dangers of haemorrhage and secondary pneumonia, which often carry off the patient within a few days of the operation.

The views of two very distinguished laryngologists on the present position of the question of radical operation for laryngeal malignant growths may be quoted with advantage. At a recent discussion at the Laryngological Society, of London, H. Butlin said that the most favourable cases for operation are those in which the disease is of intrinsic origin, and, still limited to the interior of the larynx, is of small extent, uncomplicated, and particularly in which it lies toward the front of the larynx. The more he had seen of operative surgery of malignant disease of the larynx, the more convinced he was that the removal of the whole or large parts of the larynx for malignant disease was seldom followed by sufficiently good results to justify the operation. The best results had followed and were likely to follow thyrotomy with very free removal of the soft parts in the interior of the larynx. For suitable cases such operations were comparatively free from danger. He removed Hahn's tube directly the operation was over, and made no attempt to close the wound. No tracheotomy tube was used, and no dressing in the interior of the larynx. Iodoform was frequently applied to the wound; he regarded this as of the highest importance. The patient was placed in the recumbent posture on his side with the head below. Nutrient enemata were given during the first few days, but on the day after the operation an attempt was made to give fluids by the mouth, water being tried first. He had not lost a case of thyrotomy since he had employed these measures.

Felix Semon practically agreed to everything Butlin had said, but he particularly emphasised the desirability of *early* diagnosis and *early* operation. He had seen about 100 cases in private practice; in only about 10 per cent. had he felt justified in advising a radical operation, and, broadly speaking, 50 per cent. of those operated upon had been successful, recurrence not having as yet taken place. The methods of operation selected in his cases had been (1,) partial extirpation of the larynx; (2,) sub-hyoid pharyngotomy; (3,) thyrotomy with and without resection of parts of the cartilaginous framework.

Palliative Measures.—With the supervention of respiratory obstruction, tracheotomy should be performed. Life may be prolonged many months in some cases by this operation, and in many patients there is a considerable improvement in the symptoms besides the dyspnœa. The low operation is preferable to the high, as the growth may spread down so as to render the high operation inconvenient or useless. When ulceration has occurred, the use of antiseptic applications with morphine is called for, such as aristol, iodoform, etc. ; or solutions of permanganate of potash, sanitas, etc., may be sprayed or gargled.

FIG. 56.
Rauchfuss' Insufflator for applying powders to the larynx.

The general health and strength of the patient should of course be maintained as far as possible by the exhibition of iron, quinine, and other remedies, and constipation must be combatted. The food should be soft and bland, if swallowing is painful ; it is a great mistake to urge the patient to go on eating solid food when the pain and irritation of the laryngeal disease is thereby increased.

Chapter XV.

NEUROSES OF THE LARYNX.

SENSORY NEUROSES; MOTOR NEUROSES.

SENSORY NEUROSES.

ANÆSTHESIA AND HYPERÆSTHESIA.

Etiology and Pathology.—Anæsthesia of the larynx may be *partial*, confined to the epiglottis, or to one side of the larynx, or to the supra-glottic portion; or *complete*, involving the whole of the larynx and upper part of the trachea.

It is either due to *peripheral* lesions, *e.g.*, diphtheria, syphilis, injury to the vagus or superior laryngeal nerves, or it may be *central* in origin, *e.g.*, bulbar lesions, locomotor ataxia, general paralysis of the insane, apoplexy, and possibly hysteria. Obviously it must generally be associated with motor paralyses of the laryngeal muscles, and, in many cases, with lesions of other cranial nerves.

Symptoms.—The symptoms consist mainly in a tendency for food to enter the larynx and produce spasm and choking attacks. When anæsthesia is complete and subglottic, spasm of the larynx does not occur and the food particles then set up pneumonia. Thus the prognosis is very grave if the lesion is bilateral, but Onodi has shown that the sensory nerve supply to the larynx is, to a certain extent, bilateral, and therefore complete unilateral anæsthesia is rare.

If only the superior laryngeal nerve is involved the

anæsthesia is supra-glottic, and food escaping into the trachea is more likely to be coughed out.

The *diagnosis* can only be made by probing the laryngeal mucous membrane, when the absence of sensation and spasm is readily detected.

Treatment consists in special care in feeding the patient. If bilateral, food should be given by the œsophageal tube. When due to diphtheria, local faradisation and the administration of strychnine should be resorted to, and of course in syphilis of the central nervous system, iodide of potassium and mercurial inunctions will be given.

Hyperæsthesia (and Paræsthesia) is generally due to local disease in the larynx or to hysteria. It is sometimes a marked symptom in gouty and rheumatic affections, and not infrequently a premonitory sign of tubercular disease.

MOTOR NEUROSES.

The experiments of Semon and Horsley have shown that :—

(1,) There is in each cerebral hemisphere a *bilateral* cortical centre for *adduction* of the vocal cords (as in phonation), and that in the left hemisphere this centre corresponds with the speech centre. Irritation or stimulation of either centre will produce bilateral adduction of the vocal cords, *i.e.*, spasm of the glottis ; but, of course, destruction of one centre produces no corresponding paralyses so long as the other is intact. Thus in motor aphasia, the vocal cords are not affected.

(2,) There is no cortical centre for *abduction* of the vocal cords, a purely physiological respiratory act, the reflex centre for which corresponds with the vago-accessory nucleus in the medulla.

Thus we see that it is impossible for a cortical lesion to cause any unilateral paralysis of a vocal cord, but a cortical lesion may cause bilateral spasm of the glottic muscles.

The vagus nerve, by its superior laryngeal branch, supplies sensation to the larynx on either side, and the superior and recurrent laryngeal nerves supply motor innervation to all the intrinsic muscles. But the motor fibres are ultimately derived from the spinal accessory nerve through its communication with the vagus nerve before it leaves the cranial cavity. Consequently, motor paralyses may be due to :—

(1,) Degenerative changes in the spinal accessory nuclei in the floor of the fourth ventricle ; or

(2,) Pressure on, or destruction of, the spinal accessory fibres before their junction with the vagus nerve ; or

(3,) From degeneration, injury, or pressure on the vagus trunk ; or its superior and recurrent branches ; or

(4,) The paralysis may be myopathic.

It has been conclusively demonstrated by Semon that "there exists an actual difference in the *biological* composition of the laryngeal muscles and nerve endings," rendering the abductors more prone to be affected by conditions resulting in paresis and atrophy than the adductors, "whilst the fact that also in central (bulbar) organic affections, such as tabes, the cell groups of the abductors succumb earlier than those of the adductors, points to the probability that similar differentiations exist in the nerve nuclei themselves." Thus, both in central lesions, and in peripheral lesions producing paralysis from pressure on the vagus or recurrent laryngeal nerves, we find that the abductor muscles alone are first paretic or paralysed, and that complete paralysis from subsequent involvement of the adductors supervenes after a long or short interval, unless, of course, the lesion is so gross as to produce a total and complete motor paralysis of the laryngeal nerves *ab initio*. (See Note, p. 179.)

Inspiratory Glottic Spasm.
LARYNGISMUS STRIDULUS.

Etiology.—The affection *Laryngismus Stridulus* (or "*false croup*"*) is almost invariably associated with rickets, and occurs in children from six months to two years of age, or even up to the seventh or eighth year; it also occurs in tetany.

While the remarkable excitability of the nerve centres accompanying rickety conditions is the usual predisposing cause, the spasms are often directly excited by irritation in the alimentary tracts either from undigested food or from parasites, or may be due to post-nasal adenoids, pharyngitis, or perhaps to a pendulous epiglottis (D. B. Lees).

Symptoms.—The attack consists in a sudden spasmodic closure of the glottis, with absolute cessation of respiration, lasting ten or fifteen seconds, followed by a long, crowing inspiration, either continuous or interrupted, as in sobbing. The muscular spasm may extend to the facial muscles, or the thumbs may be turned in, and carpo-pedal convulsions, or even general convulsions may supervene.

The attacks vary much in severity, and the first may be fatal from the persistence of the spasm and asphyxia, or the glottis may remain closed till unconsciousness occurs. The prognosis is especially grave in these silent cases, and in those in which the spasm is so marked that the inspiration is interrupted and sobbing. In the less severe forms the parents sometimes speak of the attacks as "passion fits," or "holding the breath" (Osler).

The attacks may occur very occasionally, or there

---

*The term "false croup," if ever employed, should be reserved for acute laryngitis in children.

may be several daily. The *prognosis* should be guarded, for the percentage of deaths is considerable.

Treatment consists in that called for in rickets or other general conditions that may be present, the removal of fæcal accumulations, parasites, etc. Warm clothing fresh air, and simple diet, and avoidance of mental excitement, are of first importance, while as regards drugs in rickets, nothing is so valuable, in my opinion, as minute doses of phosphorus.

As a rule, the attack passes off in a few seconds, but if persistent and asphyxia is threatened, the best plan is to dash cold water in the face, while the legs and body may be immersed in a hot bath. The spasm can sometimes be relieved by hooking forward the epiglottis with the forefinger.

If the attacks recur frequently, small doses of bromide of potassium, belladonna and chloral, will tend to keep them off and render them less severe.

INSPIRATORY SPASM IN ADULTS.

Etiology.—Spasm of the sphincters of the glottis in adults, is generally a *reflex* phenomenon dependent on morbid conditions of the larynx, *e.g.*, the presence of a neoplasm, tubercular disease, catarrhal conditions, or is due to irritation of the motor nerves by pressure of growths, etc. It is sometimes associated with varix of the tongue, adenoid hypertrophy of the tongue, elongated uvula, and with gouty or rheumatic laryngitis.

It occurs in certain lesions of the nerve centres, *e.g.*, the laryngeal crises of locomotor ataxia, in hydrophobia, tetany, and in hysteria. In the latter Gowers states that it may occur either in the paroxysmal, or in the rare continuous form which is accompanied with inspiratory *and expiratory* stridor.

Treatment.—This varies with the exciting cause, and

is found under their respective headings. In laryngeal crises of locomotor ataxia the inhalation of amyl nitrite gives relief, and sometimes the attacks, if slight, may be warded off by cocaine sprayed into the larynx, while other measures appropriate to the treatment of " crises " generally are demanded.

Semon believes that in the laryngeal crises of tabes dorsalis, there is an abnormal excitability of the adductor centres; it is therefore desirable to give directions to the patient to avoid, as far as possible, all sources of irritation in the larynx, *e.g.*, smoking, drinking cold fluids, etc. I have found that the removal of an elongated uvula greatly relieved the tendency to laryngeal crises in one case (see *Fig* 65).

Bidon has found digital compression of the phrenic nerve between the two inferior attachments of the sternomastoid very efficacious in arresting laryngeal spasm. The method was introduced by Leloir as a means of arresting nervous hiccough.

Expiratory Glottic Spasm.

Expiratory spasm of the glottis is due to inco-ordination of the laryngeal muscles, and occurs in a variety of conditions. Phonic spasm, and the so called "laryngeal chorea" belong to the *convulsive tics*.

PHONIC SPASM.

Phonic Spasm (*Dysphonia spastica*, Schnitzler) occurs only on attempted phonation, and is analogous to writers' cramp. B. Fränkel's cases of mogiphonia, in which spasm, with impairment or loss of voice, occurs on singing or attempts at public speaking, and in which ordinary conversation is not interfered with, come under this head.

Prosser James has described a form of phonic spasm which he terms "stammering of the vocal cords."

On attempting to phonate, the vocal cords may be seen to be so forcibly adducted that no chink is left for the expiratory current of air. The affection constitutes one form of stammering.

In the milder cases only the anterior portions of the vocal cords are approximated, and the alteration, or loss of voice, is only very transitory.

In connection with mogiphonia we may allude to another rare neurosis, although not spasmodic, the inability to whisper, "apsithyria," described by Solis-Cohen.

Treatment.—Rest, abstention for a time from the occupations with which the affection is associated, nervine tonics and the treatment of any local disease of the upper respiratory tract are the main indications. Any faulty method of producing the voice requires to be corrected by an elocutionist.

LARYNGEAL CHOREA.

This is a condition in which the glottis is spasmodically closed, and followed by a short, loud, harsh, barking cough, the "barking cough of puberty." It generally occurs in young females. The cough, which ceases during sleep, generally occurs once at a time, not a series of successive coughs (thus differing from the cough due to sensory laryngeal irritation), and recurs persistently throughout the day, even during rest. The voice is not in any way impaired, and there is no shortness of breath involving forcible inspiration after the cough. In fact it is simply a sudden closure of the glottis with a forcible expiration, due to involvement of both the laryngeal and respiratory branches of the vagus.

This affection is really one of the "convulsive tics," and not in any way associated with volitional acts. Only occasionally are general choreiform movements present in "laryngeal chorea," while in every case a particular group of muscles is involved, thus differing from the irregular variable movements of true chorea.

LARYNGEAL VERTIGO.

In this we have a condition characterised by a tickling sensation in the throat, followed by spasm of the glottis, with vertigo or momentary loss of consciousness, which is completely recovered from in a few seconds, and is not followed by stupor or any indications of the patient having had an epileptic fit. McBride explains the phenomena of laryngeal vertigo by the action of forced expiration with a closed glottis, and has shown that a similar condition of vertigo or loss of consciousness may be induced by voluntary forced expiration with a closed glottis.

If associated with a catarrhal condition of the respiratory tract, this requires treatment. In many cases the patient is otherwise healthy, and all that can be done is to administer such remedies as the bromides, and attend to the general health and hygienic conditions. It is sometimes associated with gouty manifestations.

Treatment.—Digital compression of the phrenic nerve, as recommended by Bidon for glottic spasm, might be tried in laryngeal vertigo.

PARESIS OR PARALYSIS OF THE LARYNGEAL MUSCLES.

The right and left recurrent laryngeal nerves supply all the intrinsic muscles of the larynx on their respective sides, except the crico-thyroid, the thyro-epiglottidean, and the internal thyro-arytenoidei muscles, which derive their motor innervation from the superior laryngeal nerve. The inter-arytenoid muscle is supplied by both the superior and recurrent laryngeal nerves.

This is the generally accepted motor nerve supply of

the larynx, but Exner, by a series of investigations, has shown that the motor nerve supply of the larynx may vary, and a case reported by Ruault in which the left recurrent nerve was resected (one and a half centimetres removed) for double abductor paralysis in tabes without paralysis of the adductors ensuing, seems to prove the truth of this statement.

Muscles Supplied by the Superior Laryngeal Nerve.

PARALYSIS OF THE CRICO-THYROID AND SPHINCTERS OF THE GLOTTIS.

Symptoms.—*Total paralysis* is attended by anæsthesia of the larynx, and paralysis of the thyro-epiglottic, thyro-arytenoidei interni, and the crico-thyroid muscles. The symptoms of anæsthesia of the larynx have already been referred to. The epiglottis stands erect and cannot close on the larynx, and the ary-epiglottic folds likewise do not contract on stimulating the larynx with a probe. The laryngeal appearance in bilateral paralysis of the superior laryngeal nerves is characteristic. Bosworth remarks: "I know of no lesion which will produce the curious glottis which is observed when both the superior laryngeal nerves are paralysed" (see *Fig.* 58).

I believe that in these cases the complete inward rotation of the *processi vocales* is due to the thyro-arytenoidei externi, innervated by the recurrent laryngeal nerves.

If *unilateral*, the appearance resembles that presented by simple cricoid paralysis, except that the paralysis of the thyro-arytenoid causes the flattened appearances of the normal cord to be replaced by a narrowed rounded cord. The wavy outline of the glottis, and the instability of the points of contact of the cords are likewise observed. The arytenoideus paralysis is usually not apparent, this muscle deriving sufficient innervation from the nerve of the opposite side (Bosworth).

Etiology.—If bilateral it is usually due to diphtheria or pressure of enlarged glands, but it is a rare affection.

Unilateral paralysis of the superior laryngeal nerve may be due to diphtheria, typhoid fever or hysteria, or

FIG. 57.
Unilateral paralysis of the superior laryngeal nerve. (Phonation.)

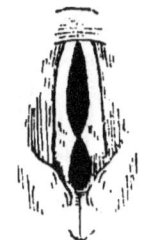

FIG. 58.
Bilateral paralysis of the superior laryngeal nerve. (Phonation.)

to injury of the nerve by pressure, or by division of, or injury to the nerve.

The treatment of the anæsthesia of the larynx has already been alluded to.

PARALYSIS OF THE THYRO-ARYTENOIDEI INTERNI.

THE INTERNAL TENSORS OF THE VOCAL CORDS.

Etiology.—This, the commonest form of myopathic laryngeal paralysis, is usually bilateral. It is generally the result of overstraining the voice, or is due to cold or catarrhal laryngitis, especially in anæmic and neurotic persons. The paralysis is then generally only partial, but affects both the internal and external thyro-arytenoidei (and is myopathic). The vocal cords cannot approximate perfectly, an elliptical space, extending throughout their length, being

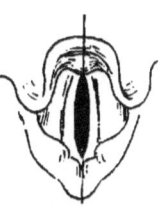

FIG. 59.
Paralysis of the thyro-arytenoidei interni and externi during vocalisation.

left on phonation, which consequently is weak, husky, or even lost. The vocal cords are practically the tendons of the thyro-arytenoidei muscles, which lie beneath the cords, and are inserted into the anterior two-thirds, and when they are paralysed the vocal cords lose their normal flat appearance, becoming rounded and narrowed.

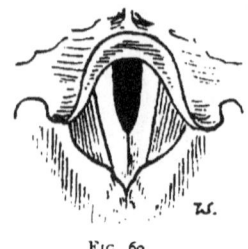

FIG. 60.
Paralysis of the thyro-arytenoidei interni in bulbar paralysis.

But while the internal muscle is innervated by the superior laryngeal, and therefore suffers in complete paralysis of that nerve, the external is supplied by the recurrent nerve. Thus it is possible in a central lesion to have paralysis of the thyro-arytenoidei interni alone, as in *Fig.* 60, the normal external muscle causing perfect inward rotation of the vocal processes.

Treatment consists in rest, the administration of general nervine tonics, and the employment of the faradic current locally. This has often to be persisted in for a considerable time, and rarely fails to cure.

Any co-existent catarrhal condition must be treated with appropriate remedies.

PARALYSIS OF THE CRICO-THYROID MUSCLE.

THE EXTERNAL TENSOR OF THE VOCAL CORDS.

The function of the crico-thyroid muscle is to render tense the vocal cords in phonation. Crico-thyroid paralysis alone is very rare. Major reports a case due to cold.* The cords are said to have a wavy margin, and if unilateral the affected cord appears higher than the other (Mackenzie). It appears to me that the wavy outline and rounded surface of the cords on expiration

---

* *New York Med. Journ.,* Feb. 20th, 1892.

and flattening on inspiration, on which Major and others have laid such stress, are due to paresis of the internal thyro-arytenoid; indeed it is difficult to see how an inflammatory lesion produces crico-thyroid paralysis without involving the other muscles innervated by the superior laryngeal nerve.

The Muscle Supplied by the Superior and Recurrent Nerves.

PARALYSIS OF THE ARYTENOIDEUS.

The arytenoideus may be paralysed alone in catarrhal conditions, from cold, or in hysteria. It is always, when apparent, bilateral, and the voice is generally lost.

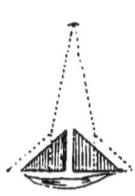

FIG. 61.
Diagram to illustrate the action of the arytenoideus.

FIG. 62.
Paralysis of the arytenoideus during vocalisation.

FIG. 63.
Combined paralysis of the arytenoideus and thyro-arytenoidei interni muscles.

A triangular chink is left between the vocal processes on attempted phonation. Local faradisation should be employed. It is combined paralysis of the thyro-arytenoid and arytenoid muscles that produces the peculiar appearance in *Fig.* 63.

Muscles Supplied by the Recurrent Laryngeal Nerves.

The abductors and adductors of the vocal cords act by rotating the triangular arytenoid cartilages on their axes.

PARALYSIS OF THE ABDUCTORS OF THE VOCAL CORDS.

The vocal cords are abducted by the crico-arytenoidei postici muscles which, arising from the posterior surface of the cricoid cartilage, pass upwards and outwards to be attached to the external angles of the arytenoid cartilages. By their contraction the arytenoids are rotated outwards on their axes, and the vocal cords are carried outwards, *i.e.*, abducted. These muscles are used in deep inspiration *(see Plate I., Fig. 3)*.

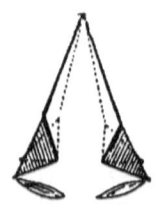

FIG. 64.
To illustrate the action of the cricoarytenoidei postici muscles in abducting the vocal cords.

Unilateral Paralysis of the Abductor is frequently met with as the result of pressure on a recurrent nerve. The reason why abductor paralysis generally occurs before complete abductor and adductor paralysis supervenes has already been discussed.

The affected cord remains in the middle line on deep inspiration being persistently retained in the position of adduction by the normal tone of the adductor muscle. Consequently, as phonation is not interfered with, and respiration is seldom much embarrassed except on exertion, the condition is frequently overlooked.

Etiology.—The most common cause is pressure on the recurrent laryngeal, either by an aneurism, tumour, enlarged mediastinal glands, tubercular thickening of the right apex of the lung, or a foreign body in the œsophagus. It is frequently due to central nerve lesions in the medulla, or from implication of the vagus or spinal accessory nerves at the base of the brain, *e.g.*, in locomotor ataxia, disseminated cerebro-spinal sclerosis, bulbar paralysis, apoplexy, syphilitic nuclear disease, or thickening of the dura mater.

It may be due to toxic neuritis, from pneumonia,

typhoid fever, diphtheria, scarlet fever, rheumatism, lead, arsenic, atropine, or may be myopathic in wasting diseases.

Bilateral Abductor Paralysis is rare, and when present is generally due to central nerve degeneration, as in the case of locomotor ataxia from which my drawing is taken. Two cases of hysterical bilateral abductor paralysis have been reported.* It may be due to cold, diphtheria, or to any of the causes producing the unilateral affection.

The cords are kept in the median position by the adductors, and only a small chink is seen between them. Consequently respiration is greatly embarrassed if the paralysis is complete, but abductor paralysis is fortunately often incomplete. Paroxysmal attacks of urgent dyspnœa with stridulous inspiration are liable to occur on slight exertion or mental excitement, and may at any time end in sudden and fatal asphyxia.

The prognosis of bilateral abductor paralysis is therefore generally exceedingly grave, and at any moment tracheotomy may be urgently demanded. In progressive lesions the adductor fibres may eventually become involved, and with the complete paralysis of the cords, which then assume the cadaveric position, respiration becomes less impeded.

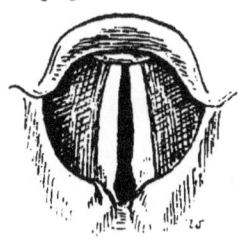

FIG. 65.
Bilateral abductor paralysis in a case of locomotor ataxia on deep inspiration.

The fact that the voice is in no way impaired accounts for the fact that bilateral abductor paralysis may exist without the slightest suspicion on the part of the patient

---

° Barling, *Brit. Med. Journ.*, April 1st, 1891; Morton, *Brit. Med. Journ.*, 1891.

or physician. In a case under Dr. Waldo at the Bristol Royal Infirmary, a newsvendor had pursued his occupation till sudden respiratory embarrassment necessitated immediate removal to the Infirmary, where, in a second attack just after admission, he died before tracheotomy could be performed. The abductor paralysis was due to a moderately enlarged thyroid gland, and the postici muscles were found completely atrophied, so that, although this patient had been the subject of double abductor paralysis for a considerable period, he had had no symptoms sufficient to attract his notice till within a day or two of his death.

Treatment.—So long as bilateral abductor paralysis is present the patient should wear an intubation or tracheotomy tube. If the condition, either bilateral or unilateral, is due to affections amenable to remedies, *e.g.*, syphilis, neuritis from diphtheria, cold, etc., they should be treated accordingly, while faradisation of the laryngeal muscles is steadily persisted in. Hysterical cases are sometimes quickly cured by the faradic current. I have found nitrite of amyl inhaled of great service in relieving the dyspnœic attack in a case complicating locomotor ataxia.

Ruault tried resection of one recurrent nerve with a view to the production of complete paralysis of one side, but without success.

PARALYSIS OF THE ADDUCTORS OF THE VOCAL CORDS.

Adduction of the cords is brought about by the action of the crico-arytenoidei laterales arising from the sides of the cricoid cartilage, and passing backwards and upwards to the external angles of the arytenoid cartilages. Their contraction causes inward rotation of the arytenoids on their axes, causing the vocal cords to approach in the middle line. But for perfect adduction of the cords the arytenoids must also be approximated by the arytenoideus and thyro-arytenoidei externi muscles. If the arytenoideus be paralysed alone, a triangular chink is left on phonation behind the closed vocal cords.

Unilateral Paralysis of the Adductor alone is extremely rare, and can only be due to local causes. It is reported to have occurred from cold, syphilis, small-pox, etc.

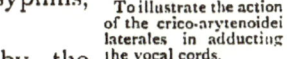

FIG. 66.
To illustrate the action of the crico-arytenoidei laterales in adducting the vocal cords.

The appearance presented by the larynx is not very characteristic, and is liable to be mistaken for complete paralysis of one vocal cord. But in total paralysis the margin is straight, while in adductor paralysis the vocal cord is completely abducted, showing a concave margin, and, the healthy cord being hardly able to completely pass across to meet it, the voice is weak and often lost.

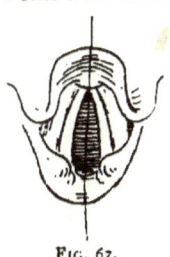

FIG. 67.
Bilateral adductor paralysis during vocalisation.

Bilateral Adductor Paralysis is generally functional, and due to hysteria. The vocal cords, on attempted phonation, are widely divergent, as in deep inspiration. In hysterical cases the voice is lost but the cough is phonic, while in those very rare cases due to local lesions, cough is aphonic, or, rather, is altogether impossible.

Laryngoplegia, or *Total Paralysis* of the vocal cord, is the usual result of pressure on the recurrent laryngeal nerve. If bilateral, the vocal cords remain in the cadaveric position, both during phonation and respiration.

In *laryngo-hemiplegia,* or total paralysis of one vocal cord (*see Fig.* 68), the respiration is hardly affected. The voice is generally lower in pitch than normal, is readily tired, and often hoarseness is observed, but ordinary conversation is possible. In male patients, at any rate, paralysis of one cord is often never suspected, because it is erroneously assumed that *aphonia* must result from paralysis of a vocal cord.

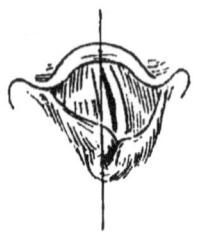 

FIG. 68.
Paralysis of the left vocal cord during phonation.

FIG. 69.
The same on deep inspiration.

During quiet respiration the larynx appears normal, but in vocalisation the healthy cord is over-adducted, and passes across the middle line to meet the paralysed cord, producing a peculiar distortion of the laryngeal image; the arytenoid cartilage on the paralysed side being unsupported by its muscles is *pushed aside*, and lies behind the sound and over-adducted arytenoid, and, like the corresponding vocal cord, appears to lie at a somewhat lower level than on the sound side.

On deep inspiration, as shown in *Fig.* 69, the arytenoid on the healthy side passes further *back*, as well as being more abducted than on the paralysed side.

The significance of unilateral paralysis of the vocal cord is of considerable importance in medical practice, but presents many difficulties to those who have had no opportunity of special study of the larynx.

Unilateral paralysis of the *left* cord may be due to	Cancer of the œsophagus Aneurism of the aorta Goître Syphilis	} Frequent
	Ataxia Hysteria Cold	} Less frequent
	Bulbar lesions Cerebral lesions Primary neuritis of the recurrent	} Rare
Unilateral paralysis of the *right* cord may be due to	Aneurism (of the arch, subclavian) Cancer of the œsophagus Goître Various tubercular lesions Syphilis	} Frequent
	Ataxia Cold	} Less frequent
	Bulbar lesions Cerebral lesions Primary neuritis	} Very rare *

Unilateral paralysis of either cord may be simulated by disease of the crico-arytenoid joint leading to fixation—a rare occurrence.

ANKYLOSIS OF THE CRICO-ARYTENOID JOINT.

Although unconnected with neuroses of the larynx, it will nevertheless be convenient to refer here to diseases of the crico-arytenoid joint, which lead to fixation of the vocal cords, and which in their clinical aspect therefore, constitute one form of vocal cord paralysis.

Any degree of stiffness of the crico-arytenoid joint is included under the term "ankylosis," whether it be due to disease of the joint itself (" true ankylosis") or to disease in the immediate neighbourhood of the joint which mechanically interferes with the movements of the arytenoid cartilage (" false ankylosis").

Of course, all acute inflammatory affections involving the arytenoid cartilage, or the mucous membrane

---

* Altered from Morgagni.

covering it, are very likely to involve the crico-arytenoid joint, or at least to cause fixation of the arytenoid cartilage from inflammatory exudation, as, for instance, in gouty and rheumatic attacks, the exanthemata, diphtheria, and from mechanical injury, while cancer, syphilis, and tubercle may be cited as frequent causes of less acute ankylosis, with or without arthritis.

When the crico-arytenoid articulation is ankylosed on one side only, the laryngoscopic appearance very closely resembles laryngo-hemiplegia, but in unilateral recurrent paralysis, as we have already seen, the arytenoid cartilage on the paralysed side is displaced by the sound and over-adducted arytenoid cartilage, whereas in ankylosis this is not the case.

Attention to this point sometimes enables one to distinguish between a paralysed vocal cord and simple fixation of the cord from disease implicating the crico-arytenoid joint, for in the latter case the healthy arytenoid approaches but *does not displace* that on the

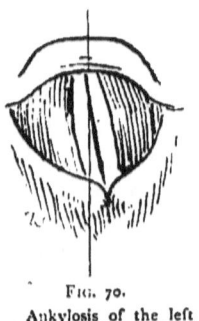

Fig. 70.
Ankylosis of the left crico-arytenoid joint during vocalisation.

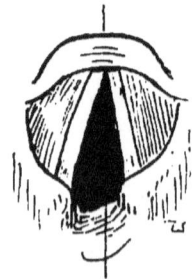

Fig. 71.
The same on deep inspiration.

affected side. This fact led me to diagnose simple arytenoid fixation in a patient that was said to have a thoracic aneurism with resulting abductor paralysis, and whose laryngeal condition is shown in *Fig.* 70. The absence of aneurism, after the laryngeal symptoms had existed for two years, was confirmed by two other physicians. Yet, in course of time, a simply paralysed cord becomes more or less fixed from disuse.

This patient had noticed huskiness of the voice two years before I saw him. His only other illness was diphtheria seventeen years previously, with paralytic sequelæ. I have not seen him for two years, but he is reported perfectly strong and well.

Diagnosis.—Attention has been directed to one useful point in the differential diagnosis of ankylosis of the joint.

Bilateral ankylosis is most liable to give rise to error in diagnosis, resembling bilateral recurrent paralysis, but complete paralysis of both vocal cords (apart from ankylosis) is extremely rare. In many cases the swelling in the arytenoid region indicates the cause of the fixation of the cord.

LUXATION ON THE CRICO-ARYTENOID JOINT.

This is an extremely rare affection. Cheval reported a case in *La Clinique*, xv., 1891, which he was able to reduce by means of a strong faradic current, a double electrode being applied to the posterior wall of the larynx so as to tetanise the inter-arytenoid and posterior crico-arytenoid muscles.

THE LARYNGEAL MANIFESTATIONS OF CHRONIC DISEASES OF THE CENTRAL NERVOUS SYSTEM.

It is to the labours of Semon and Burger that we are chiefly indebted for our knowledge of the laryngeal affections in *locomotor ataxia*, in which there are four kinds of laryngeal affection that may occur: (1,) *Sensory disturbances*, anæsthesia, hyperæsthesia, paræsthesia. Various abnormal sensations in the larynx precede or accompany laryngeal crises, and in a few cases persistent anæsthesia has been reported. (2,) *Inco-ordination* of the laryngeal muscles, or laryngeal ataxy. The voice may be jerky, or resemble dysphonia spastica in being interrupted by intervals of aphonia, or huskiness. It is a rare affection, and only one case have I seen in which the abductor and adductor movements of the cords were ataxic, and the symptoms were not marked. (3,) *Laryngeal crises* are frequently present in locomotor ataxia. According to Gowers they are almost as frequent as gastric crises. They may constitute the earliest manifestation of locomotor ataxia, and in a large proportion of cases are associated with abductor paresis. The patient feels a sense of tickling or dryness, or of stricture in the laryngeal region, quickly followed by a succession of abrupt coughs—resembling whooping-cough—continuing till the patient feels almost asphyxiated, and being followed by inability to inspire, or by a long-drawn whoop, during which air is drawn into the chest with very great difficulty. The whole attack may last but a quarter of a minute, or may persist for five or ten minutes. Death from asphyxia is unusual, but has occurred. In some cases the laryngeal crisis is attended by loss of consciousness, vomiting, vertigo, or pains in the chest or limbs. (4,) *Paralysis* of the abductors and adductors of the cords. Burger collected seventy-one cases recorded up to the year 1891, and in forty-three of these there was either unilateral or bilateral paralysis of the abductors. The symptoms of abductor paralysis are described on p. 168. After abductor paresis has lasted for some time, it may be followed by the supervention of adductor paralysis.

In one of my patients (*Fig. 65*) the pulse rate was persistently frequent. Saundby and Semon have drawn attention to this feature in abductor paralysis in tabes.

It is due to the fact that the inhibitory nerve of the heart is, like the motor nerves of the larynx, derived from the accessory nucleus, and thus I hold that *paresis of the abductor muscles, associated with persistent increased pulse frequency*, should always lead to the suspicion of tabes.

At the Bristol meeting of the British Medical Association, 1894, Dr Permewan communicated the result of the laryngoscopic examination of thirty-four cases of general paralysis of the insane, and concluded that in at least 20 per cent. of these cases there was more or less marked abduction paresis. In one of his cases there was paresis of the abductor of one cord, associated with complete abductor paralysis. His observations confirmed the general truth of Semon's law as to the particular susceptibility of the abductors to succumb to organic disease. It is interesting to note that in Permewan's cases there was no regular combination of symptoms pointing to the existence of posterior sclerosis, although there is undoubtedly such a close relation between general paralysis and tabes dorsalis.

In *labio-glosso-laryngeal paralysis*, anæsthesia of the larynx has been noted by Kussmaul. Laryngeal crises are almost unknown in this affection, but there are several recorded cases of abductor paresis. I have met with one instance (*Fig.* 60) in which apparently thyro-arytenoid paralysis was present alone, without any abductor paresis, although the other usual features in the tongue and soft palate were well marked. Permewan in the paper referred to above, related his history of a case in which paresis of the adductors was observed with complete bilateral abductor paralysis within nine months of the commencement of abductor paresis.

In *disseminated sclerosis*, laryngeal paralysis is very rare, but the vocal cords in one case under my care exhibited fine irregular tremor during phonation. The slow, monotonous tone, with jerky voice and scanning speech, is an early feature in most cases.

DIAGNOSIS OF CASE OF INTRA-LARYNGEAL PARALYSIS.

The differential diagnosis of paralysis of the laryngeal muscles often presents considerable difficulty to practitioners, and the following case is introduced for the purpose of drawing attention to the method by which the diagnosis may be made, and as also showing the value of laryngoscopy as an aid to the diagnosis of some general affections.

Mr. X., Taunton, age 33, complained of having had a constant cough, short, and coming on in paroxysms at frequent intervals, and frequently ending in vomiting. This had been going on for many months, and he was wasting and felt weak. Pulse rate 108. There was no evidence of lung disease, and the heart sounds were clear and normal. There was no history of syphilis, alcohol, etc.

The nose was examined; touching the ·posterior extremities of the middle turbinals readily excited cough, but the nasal passages were normal in appearance.

The larynx appeared to be perfectly healthy in colour and contour, and on phonation the vocal cords were apposed in the middle line, the voice being unaffected. But on inspiration, the right vocal cord remained in the middle line, while the abduction of the left was variable and imperfect.

There was no indication of any inflammatory deposits that might have led to the suspicion of fixation of the right crico-arytenoid joint. Nor was there any recent history of diphtheria, alcoholism, etc., that could have produced a peripheral paralysis. Was it due to pressure on the right recurrent laryngeal nerve? Careful examination of the chest revealed no indication of any intra-thoracic tumour or aneurism. The apex of the heart was not displaced. The pulse, though frequent, was equal on both sides. The fact that abductor paresis of one side was combined with the abductor paralysis on the other was strongly suggestive of a lesion of the central nervous system (bulbar), and the frequent pulse, in the absence of other affections to account for it, left little doubt in my mind that the vago-accessory nuclei were extensively involved. On questioning the patient, it appeared that the paroxysms of coughing were often followed by a long inspiratory "whoop," and attacks of vertigo. On asking him to stand with his eyes closed, he exhibited well-marked ataxic symptoms. The patellar tendon reflexes were absent. The stomach was dilated with a large quantity of fluid contents. There were no other symptoms or signs of locomotor ataxia; the Argyll-Robertson phenomena, alterations in the optic disc, lightning pains, paræsthesia in the limbs, etc., were absent, but there could be little doubt that the paroxysms of cough were due to "laryngeal crises" of tabes. In one of the cases reported by Semon, abductor paresis was discovered two years before the other symptoms of tabes appeared, while there are many cases recorded in which the laryngeal symptoms preceded the appearance of typical locomotor ataxia by one year. Possibly the dilatation of the stomach in Mr. "X.'s" case was due to a paresis and wasting of the muscular walls, analogous to the condition of the posticus muscles in the larynx.

Case of Laryngeal Paralysis following section of the vagus nerve.—A remarkable proof of the truth of Semon's law of the proclivity of the *postici* muscles (the abductors of the vocal cords) to succumb to organic disease, is afforded by the case of A. P., a boy aged 15, under the care of one of my surgical colleagues, at the Bristol Royal Infirmary. In performing an operation for septic thrombosis in the right internal jugular vein, under conditions which rendered the procedure especially difficult, the right vagus nerve was divided below the superior laryngeal branch, the cut ends being sutured immediately. When I examined the larynx shortly afterwards, the right cord was completely paralysed, and it remained in this condition for several months. But on examining the larynx in July, 1894, seven months after the operation, the right cord was in the middle line, and the boy had an excellent and powerful voice. Thus there was no longer complete paralysis, but only abductor paralysis. The patient was seen by Dr. Semon and Dr. McBride, both pronouncing it to be a case of simple abductor paralysis of the right cord.

Thus, not only are the *postici* muscles more prone to succumb to organic disease than the *laterales* (or adductors), but, as this case shows, the adductor nerve fibres tend to regenerate and recover functional activity sooner than the abductor fibres.

Chapter XVI.

FOREIGN BODIES IN THE LARYNX.

It is unnecessary to enumerate the many varieties of foreign bodies that have been found in the larynx. Pieces of fish bone, or meat bone, or pins are commonly impacted in the larnyx, and from their shape tend to "stick."

Coins and flat bodies may lie across the glottis, with the margins in the ventricles, as in the case I have illustrated. For removing foreign bodies in this position Wolfenden's or Cuzco's forceps are particularly well adapted.

Fig. 72.

Schliep has found that the application of vinegar will

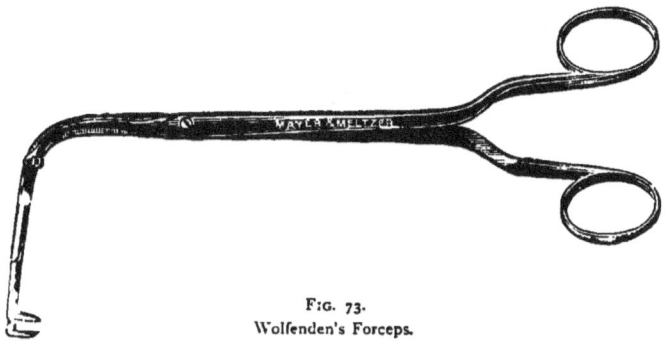

Fig. 73.
Wolfenden's Forceps.

soften fish bones in fifteen or twenty minutes, while a 1 to 5 per cent. solution of hydrochloric acid, applied repeatedly by a cotton wool tampon, will soften meat bones. If situated in the œsophagus, small quantities of this solvent may be repeatedly swallowed.

We need not discuss the obvious symptoms that follow the entrance of a foreign body into the larynx, but it may be remarked that, after the first glottic spasms, the larynx soon learns to tolerate its presence to some extent. Dyspnœa and cough may be due to the body having passed into a bronchus.

If, on laryngoscopic examination, the body can be seen, it can generally be removed by suitable forceps, but the greatest care should be observed to avoid pushing it into the trachea.

It is often a very difficult matter to detect small bodies, such as pieces of fish bone, which may get stuck in the lateral ary-epiglottic angle, being almost concealed from view on laryngoscopy, and it is well to bear in mind that the sensations of the patient are rarely any guide to the situation of a foreign body in the larynx. A body in the glosso-epiglottic space or even in the naso-pharynx, is referred subjectively to the larynx, and thus the sensations of the patient may lead us to overlook a foreign body which has lodged here.

The soreness and irritation caused by a fish bone or other body, will often cause a patient to believe it is still in the throat, for some hours after it has been dislodged.

Chapter XVII.

RHINOSCOPY.

ANTERIOR RHINOSCOPY; POSTERIOR RHINOSCOPY.

RHINOSCOPY.

By referring to the *Frontispiece* the actual anatomical relations and form of the parts now to be described may be compared with their appearance, when, as seen during life by anterior and posterior rhinoscopy, they can only be observed "end on."

It is unnecessary to refer again to the subject of lighting apparatus, but a brilliant illumination is even more essential in the examination of the nasal passages than in laryngoscopy, while the same forehead mirror of about fourteen-inch focus should be used.

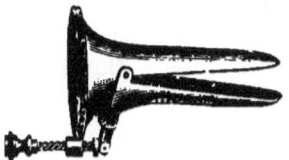

FIG. 74.
Duplay's modification of Bresgen's Speculum.

FIG. 75.
Lennox Browne's Nasal Speculum.

Anterior Rhinoscopy.—For this we require a nasal speculum. Fränkel's is simple, but the vibrissæ are apt to project through the fenestræ of the blades and obscure the parts beyond, and therefore I think that for general use one of the best is Lennox Browne's, consisting of two ivory blades on sliding bars, as it may be used in cautery operations as well as for examination purposes. The spring of Thudicum's is liable to cause pain, unless very carefully held. The most comfortable to the patient is Bresgen's, but it is more difficult to manipulate. Some form of self-retaining speculum, like Cresswell Baber's or Neil-Griffiths's, is desirable for operations on the nose

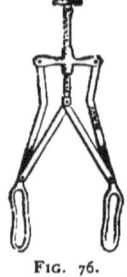

FIG. 76.
Fränkel's Nasal Speculum.

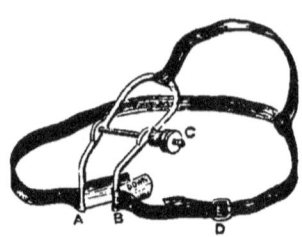

FIG. 77.
Neil-Griffith's Self-retaining Speculum.

requiring two hands. A few fine silver probes, to investigate the nature and consistency of the various prominences and swellings encountered, complete the apparatus for anterior rhinoscopy.

Having previously noticed any indication of nasal obstruction, and tested each nostril separately, let the light be focussed on the anterior nares. But before introducing the nasal speculum, raise the tip of the patient's nose and examine the vestibule and the front part of the nasal passages, otherwise ulcers, septal perforations, or other abnormalities may be subsequently concealed by the blades of the speculum and escape detection.

Then, with the patient's head erect, insert the speculum and gently separate the blades, directing the light well into the passage. Observe the olfactory fissure, and note its width and whether there are any collections of abnormal secretion. The inner wall is formed by the septum, covered with yellowish pink mucous membrane; any departure from the normal contour, the presence of ulceration, new growths, septal deviations or perforation will be easily recognised here.

On inspecting the outer wall, the inward projection of the inferior turbinated body first arrests attention. Even when healthy it varies considerably in colour and size, according to the state of the erectile tissue; for when the venous sinuses are distended the turbinal body appears as a full, red or pink, tense swelling, as shown in *Plate V., Fig 1;* but when collapsed it is pale pink, and is much less prominent. The turbinals are shown in this figure in a state of abnormal distension, as in vaso-motor rhinitis.

By directing the patient to throw back the head, we bring into view the anterior portion of the middle turbinated body, which is similar to the inferior in colour and consistence. The presence of polypi in the middle meatus, or any discharge of pus here, should lead to a careful investigation of the region of the *hiatus*. Seldom can we see more than the anterior border of the superior turbinated body. Sometimes it is possible to see the posterior pharyngeal wall through the normal nasal passages, especially when the olfactory fissure is wide, and then, on swallowing, the salpingo-palatine fold is seen crossing inwards.

FIG. 78.
Zaufal's Funnel Speculum for examining the posterior wall of the naso-pharynx.

With a probe the consistence of any growth may be investigated, and whether it is fixed or moveable; vascular engorgement of the turbinal bodies gives the sensation of a cyst distended with fluid, and pits on pressure. By spraying the parts with a 5 per cent. solution of cocaine, the vessels are emptied, and the swelling due to simple vascular distension disappears, and thus not only do we get a clearer view of the parts beyond, but we can determine how much of any swelling is due to vascular engorgement, and how much to true hyperplasia of the tissues.

Posterior Rhinoscopy is more difficult of accomplishment, and requires considerable practice before a satisfactory examination can be made. One of the smaller laryngeal mirrors (half inch diameter) may be used for the purpose, but a special form of rhinoscopic mirror, such as Fränkel's or Michel's, with moveable mirrors, is much easier to manipulate.

FIG. 79.
Michel's Rhinoscope.

Direct the patient to open the mouth, with the head erect, then with the left hand depress the tongue; for this purpose Türck's depressor is the most convenient, as the twist in the handle enables the left hand to be kept well out of the way of the right holding the rhinoscope. In using it care must be taken to place it just beyond the dorsum of the tongue, and no further; for if not far enough back the tongue bulges up and occludes the view, while if too far back gagging and retching are induced.

Let the patient breathe quietly, and generally the soft palate will very soon relax spontaneously. If it does not do so, instruct the patient to *breathe in* through the nose. If this manœuvre fails, we must use a palate retractor to draw forward the soft palate. Voltolini's or

White's self-retaining retractors are convenient forms. Before using it, the fauces and the back of the soft palate should be cocainised, and sometimes the application of cocaine obviates the necessity for the retractor. As soon as the soft palate ceases to occlude the nasopharynx, the rhinoscope should be introduced and

Fig. 80.
Voltolini's Palate Retractor.

passed below and well behind the velum, care being taken to avoid touching the fauces or pharyngeal wall.

By gradually tilting forward the mirror, the posterior border of the septum nasi is brought into view in the centre of the image *(see Plate V., Fig. 2)*. Then on slightly turning the mirror, first to one side, then to the other, the *choanæ* or posterior openings of the nasal

Fig. 81.
Cresswell Baber's modification of White's self-retaining Palate Retractor.

passages come into the field, and, projecting inward, the posterior extremities of the inferior and middle turbinal bodies appear as " frogspawn-like," grayish-white bodies. Owing to the very oblique direction in which the parts are viewed in rhinoscopy, the lower turbinals appear to rest on the upper surface of the soft palate, the inferior meatus being seldom seen at all. Sometimes the superior turbinal can be dimly seen in the upper part of the choanæ. Below, the superior or posterior surface of the soft palate and uvula come into view, and, turning the mirror to either side, we can see the orifices of the

Eustachian tubes, and behind them the fossæ of Rosenmüller. Finally, after once more finding the posterior border of the septum, which acts as a convenient land mark, the roof and posterior wall of the naso-pharynx are successively brought into view, as the mirror becomes more and more horizontal before it is withdrawn.

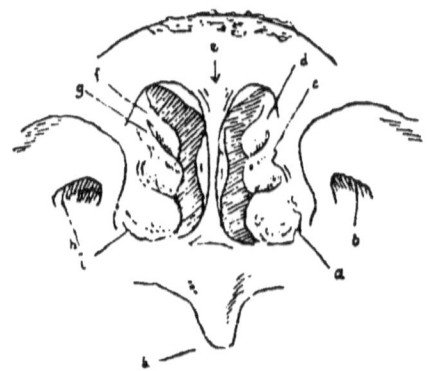

Patient's Right.   Patient's Left.
FIG. 82.

Explanatory Diagram of the Rhinoscopic Image, *Plate V.*, *Fig.* 2.   *e*, the septum, on either of which are the *choanæ narium* ; *a, c, d,* the left inferior, middle and superior turbinated bodies ; *i, e, f,* the right ditto ; *b* and *h,* the left and right Eustachian tubes ; *k* posterior surface of the uvula.

Of course only small portions of the naso-pharynx can be seen in the rhinoscopic mirror at any one time, but by mentally connecting the several successive images we obtain a complete rhinoscopic image as depicted in the plate.

While making the examination, the presence of any growth or collections of secretion, or other abnormality, should be carefully noted.

It is scarcely necessary to add that the rhinoscopic examination must be very rapid, and very often we have to make several short successive examinations, or even be satisfied with such a momentary glance that only the practised eye could see the condition of the parts.

PLATE V.

The Nose and Naso-Pharynx.

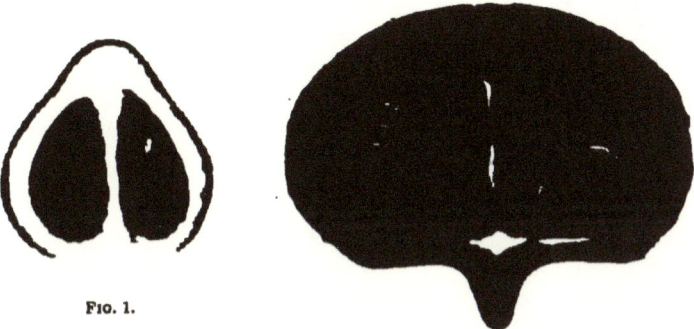

Fig. 1.

Fig. 2.

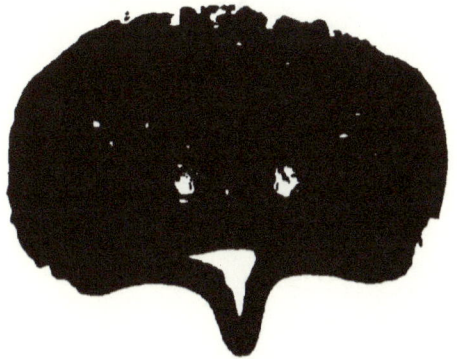

Fig. 3.

Fig. 1.—The anterior nares.
Fig. 2.—The posterior rhinoscopic image in a healthy adult.
Fig. 3.—The posterior rhinoscopic image in an adult showing adenoid growths in the vault of the naso-pharynx, and hypertrophy of the inferior turbinated body.

CONGENITAL DEFECTS AND DEFORMITIES.

We need not here refer to congenital defects or absence of the nose externally, conditions which come under the scope of general surgical text books.

It is not very unusual to find the anterior orifice of the nasal passages partially obstructed by a web-like band projecting up from the floor of the vestibule about half an inch from the margins of the alæ. Very rarely this web of skin completely occludes the nasal passages, representing the persistence of a condition analogous to the congenital web sometimes observed partially occluding the glottis. I have seen one such case, another is recorded by Jarvis.

A congenital osseous occlusion of the posterior nares, either unilateral or bilateral, has been observed, but such conditions are extremely rare.

The most frequently occurring congenital defect is deviation of the septum. It is, in fact, unusual to find a septum which is perfectly straight. The condition is referred to more fully under the head of deviated septum. The septum has been found to be doubled (splitting of the septum), and very rarely the septum is continued back into the naso-pharynx, which is then divided into two chambers, more or less completely.

Chapter XVIII.

RHINITIS.

ACUTE RHINITIS, FIBRINOUS RHINITIS, CHRONIC RHINITIS, HYPERTROPHIC RHINITIS, ATROPHIC RHINITIS, RHINITIS ÆDEMATOSA AND CASEOSA.

ACUTE RHINITIS.

Simple Acute Rhinitis, the common cold in the head, hardly requires lengthy notice here. There is good reason to believe that it is due to a micro-organism affecting a susceptible mucous membrane, having an incubation period of about two days, and which, in individuals predisposed, acquires increased virulence; thus it is we often see one weakly member of a family who is continually catching colds in the head and infecting the other less susceptible members in turn.

At any rate, practical experience has shown that it is possible to abort a cold if it be taken in time, and most of the remedies that have proved successful are local stimulants to the nasal and pharyngeal mucosa or local germicides. Thus Lederman, of New York, has found the following mode of treatment very beneficial during the congestive stage of an acute nasal catarrh. The nasal chambers are sprayed with any of the antiseptic solutions, Seiler's preferred, until they are sufficiently cleansed, and then the following solution is used :—

℞ Cocaine
    Menthol āā   - - - -   grs. xx
    Benzoinol - - - -   f℥ij
·M. ft. solutio.

Capitan has recommended the frequent insufflation of the following :—

℞ Pulv. talc	-	-	-	-	grs. lxxv
Antipyrin	-	-	-	-	grs. xv
Acid boric (pulv.)	-		-	-	grs. xxx
Acid salicylic	-'		-	-	grs. iv

I have frequently prescribed Formula 44 with good effect.

Attention to the general health and general hygienic surroundings, exercise in the open air, and the daily use of the cold bath, or cold bathing, will do more than any local treatment towards rendering the patient more resistent and less liable to colds. Acute rhinitis may be simply the result of a chill, a vaso-motor rhinitis, the symptoms of which come on in a few hours after exposure, and are followed by a simple catarrh which may pass off in a day or two, or may often be checked effectually by a Turkish bath, or a hot bath and a Dover powder on going to bed.

The *acute purulent rhinitis* of young children is generally due to infection at birth by leucorrhœa or gonorrhœa of the maternal passages, and, of course, requires very careful treatment with cleansing lotions and mild antiseptics.

CHRONIC RHINITIS.

Etiology.—Simple chronic rhinitis may result from frequently recurring attacks of acute rhinitis, which results in a certain degree of permanent thickening of the tissues and chronic congestion—a more or less persistent chronic rhinitis with frequent slight or acute exacerbations. Chronic rhinitis is thus often left in children after measles or scarlet fever, and is associated with more or less ill health, a condition which should not be neglected or treated too lightly, as it so often leads to the implication of the middle ear. Doubtless many cases of otitis media in children could be prevented by timely treatment of chronic rhinitis.

In adults we find other causes of chronic rhinitis in the inhalation of irritating particles of dust, such as are produced by working in stone, mattress making, and upholstery. In children a form of *chronic purulent rhinitis* is sometimes associated with the strumous diathesis.

Rhinitis sicca is a chronic rhinitis attended with deficient secretion, the mucus becoming inspissated and tenacious, or forming simply dry crusts of mucus unattended with fœtor. It is found chiefly in anæmic girls, or as a manifestation of the gouty or rheumatic diathesis, or may be due to chronic alcoholism. It is often associated with pharyngitis and laryngitis sicca.

Symptoms.—The only marked symptom is constant nasal discharge; in elderly patients especially this may be simply a copious watery exudation. The nasal mucous membrane is sometimes congested and the turgid turbinal bodies pit on being touched with a probe. These cases should perhaps be regarded rather as a form of vaso-motor rhinitis. When the nasal discharge is due to irritation of particles of dust or, as in children, is a sequence of the exanthemata, the discharge is usually muco-purulent and strings of sticky mucus occupy the nasal passages. Sajous mentions as a common symptom, pain over the brow, coupled with a feeling of weight, due to inflammatory narrowing of the infundibulum.

Treatment.—*General.*—In dealing with the chronic forms of rhinitis, attention to the general health is, of course, essential to success. It is too often regarded as a purely local affection. Indigestion, a tendency to constipation or torpidity of liver, if present, should be treated. The possible existence of post nasal-adenoids should be borne in mind. Change of air or a voyage, out-door exercise, avoidance of hot, overcrowded rooms and late hours, will often succeed when local treatment alone has failed; and when we remember that chronic rhinitis is often largely dependent on disturbance of the neuro-vascular mechanism, that it is sometimes almost a neurosis, the necessity for a generous tonic line of treatment is obvious. Massage and cold douching daily are most beneficial, and I may cite the case of a medical friend of mine who having suffered from chronic rhinitis for years, and in whose case local treatment had been attended with no lasting improvement, underwent a course of Turkish baths and massage continued for several months in succession, and has remained perfectly free from his nasal trouble for many years. Many similar instances have occurred in my experience.

Beverley Robinson recommends a tablet containing gr. ¼ each of chloride of ammonium and powdered cubeb with some liquorice, together with codeine if there is much cough, taken every fifteen or thirty minutes, or every hour for some time.

Local.—In the earlier stages of chronic rhinitis, such general measures may be aided by mild local applications. I have found the greatest benefit from liquid vaseline containing terebene (♏ x ad ℨj), eucalyptol (grs. xv—xx ad ℨj), with camphor (gr. i—ii ad ℨj) sprayed well into the nose with an oil atomiser. This should be done night and morning for some weeks. If crusts of dried secretion tend to collect in the passages, a weak solution of bicarbonate of soda and borax in warm water should be forcibly sprayed in for some minutes till they are loosened and can be cleared out, the oily solution being subsequently used. A post-nasal spray must be used occasionally by the medical attendant, to ensure thorough removal of all collections of inspissated secretions in the naso-pharynx.

As the condition improves, and the mucous membrane of the nose becomes more tolerant, a snuff composed of sodium chloride ʒij, boracic acid ʒss, ammonium chloride ʒss, camphor gr. j, may be used twice daily in the place of the other local remedies.

These are the remedies I have found most generally useful, but many others are recommended by various authorities, such as solutions of zinc sulphate grs. ij, alum grs. iv to viij, zinc chloride gr. j, nitrate of silver, grs. ij to xv, or sodium benzoate grs. xxx to the ounce; tar water; insufflations containing nitrate of silver, tannic acid, iodol, sozoiodol, potassium sozoiodolate, sanguinaria, etc.

CHRONIC HYPERTROPHIC RHINITIS.

Etiology.—This may be regarded as an advanced stage of simple chronic rhinitis, and may, therefore, be due to those conditions which lead to chronic rhinitis. From the histological investigation of twenty cases Wyatt Wingrave found almost invariably mucoid degeneration of the muscular walls of the venous sinuses, and he suggests that in the persistent distension of the sinuses which ensues we have the explanation of the general hyperplasia associated with "turbinal varix."

Symptoms.—The nasal obstruction resulting from the hypertrophy of the tissues, and the constant presence of thick tenacious mucus which passes into the naso-pharynx and leads to constant hawking, are very distressing to the patient. The symptoms, in fact, are those of simple chronic rhinitis greatly intensified, the obstruction of the nasal passages being often so considerable that the patient cannot blow his nose properly.

Particularly in this form of rhinitis are reflex nasal neuroses prone to occur, especially a persistent, hard, spasmodic cough, analogous to that cough which is excited by the passage of the Eustachian catheter. Asthma and hay fever are likewise often due to the existence of this condition in the nose.

On *examination* of the nasal passages, we find a general thickening of the mucous membrane. Especially is this noticeable in regard to the lower turbinals; these become greatly thickened and tough from the overgrowth of fibrous trabeculæ between the venous spaces, which are more or less obliterated. The olfactory fissure is often nearly blocked by these hypertrophied inferior and middle turbinated bodies.

With the rhinoscopic mirror the inferior turbinals are seen to have become even more hypertrophied posteriorly, the pale grayish-white, mulberry-like extremities more or less blocking the choanæ (see *Plate V., Fig. 3*). Mucopurulent secretion is always present, and must be washed off by a coarse alkaline spray, in order to see clearly the condition of the nasal structures. These hypertrophied turbinals do not collapse on applying cocaine. The cocaine spray thus affords a ready means of distinguishing hyperplastic overgrowth from simple venous distension of these normally vascular structures.

There should be no difficulty in distinguishing the pale-gray fibrous-looking hypertrophied turbinal bodies from a mucous polypus, which is moveable, soft, and semi-translucent. In addition to the abnormal conditions already mentioned, we very often find a narrowing of the nasal passages from a deviation of the septum, or a spur, or some other structural deformity which has had an active share in setting up the conditions which lead to hyperplastic changes in the nasal mucosa.

Treatment.—When we have to deal with the hypertrophic form of rhinitis, no local remedy can compare with the use of the galvano-cautery, or the snare. In well marked cases, it is hopeless to expect a reduction of the mass of hypertrophied tissue by other methods. By the application of a 10 per cent. solution of cocaine, we not only

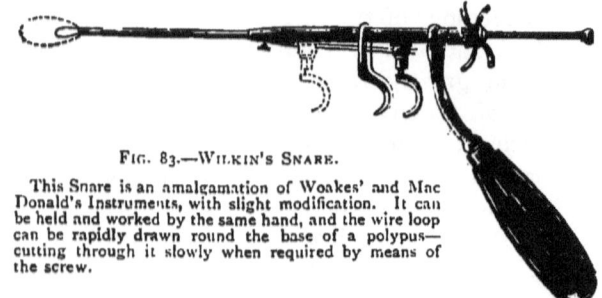

FIG. 83.—WILKIN'S SNARE.

This Snare is an amalgamation of Woakes' and Mac Donald's Instruments, with slight modification. It can be held and worked by the same hand, and the wire loop can be rapidly drawn round the base of a polypus—cutting through it slowly when required by means of the screw.

render the parts insensitive, but are able to judge how much of the enlargement is due to fibro-plastic new growth which requires removal. If the turbinal hypertrophy can be secured in a snare, it may be removed by this means, the result obtained by this instrument proving highly satisfactory to those who adopt it; and if the cold snare be very gradually tightened our object can be accomplished with little hæmorrhage, although for this purpose the galvano-caustic snare is preferable, as it cuts through the tissues more rapidly. It is often extremely difficult to engage these masses of hypertrophied turbinal tissue and in many cases it is necessary to first transfix the mass with a Jarvis needle to prevent

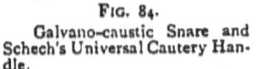

FIG. 84.
Galvano-caustic Snare and Schech's Universal Cautery Handle.

the wire of the snare slipping off as the loop is tightened.

When the hypertrophy is not suitable for snaring, one deep linear cautery groove should be made along the whole length of the inferior turbinated bodies. Two or three of these cautery incisions are usually necessary to reduce the hypertrophy, and at least one week should be allowed to elapse between each.

Wyatt Wingrave, at the meeting of the British Medical Association at Bristol, 1894, communicated the result of his investigations based on 200 cases operated on during the preceding three years at the Central London Hospital, and advocated the use of Carmalt Jones' ring knife in cases presenting very marked turbinal varix.

In slighter cases Spitzer commends the application of solution of iodine (see p. 36.); but I generally find that the application of pure trichloracetic acid, or chromic acid, to the previously cocainised hypertrophic tissue, proves most successful. These powerful caustics must only be applied on small areas by means of a probe carefully guided by inspection.

As in using the cautery, the part to be acted on by the acid should be carefully dried by a probe carrying a pledget of absorbent wool. The application should then be made with a caustic carrier, or a flat probe, and care should be taken to ensure that a small quantity of the caustic lies on one side of the carrier only, otherwise it will spread over a considerable area. The probe carrying the caustic agent should be firmly held in contact with the part to be cauterised for ten or fifteen seconds; on removing it, the white eschar is plainly visible. It is well to spray the part with a little cocaine solution, or a warm alkaline lotion, so as to wash off any remains of acid. The process generally requires to be repeated a few times, a sufficient interval being allowed between each application for the superficial eschar to come away, a mild antiseptic and alkaline lotion being used as a spray in the intervals.

General Treatment is, however, quite as necessary in this affection as in simple chronic rhinitis, and all that has been said under this head in reference to simple chronic rhinitis applies equally to the more advanced hypertrophic form with this difference, that local treatment must occupy the first place in the latter disease, whereas it is comparatively unimportant compared with general therapeusis in the early simple rhinitis.

ATROPHIC RHINITIS.

Etiology and Pathology.—There is considerable diversity of opinion as to pathology of atrophic rhinitis. Thus: (1,) Zaufal believes ozæna arises from a congenital deficiency of the turbinated bones, resulting in undue patency of the nasal passages. It is difficult to accept such a view, for, though a disease of youth, it is never congenital, and, moreover, may undergo spontaneous cure; (2,) The opposite view is held by Berliner, that it is associated with nasal obstruction, and due to pressure of the middle turbinal against the septum, with consequent defective secretion; (3,) Habermann extending B. Fränkel's and Krause's observations, regards ozæna as due to fatty degeneration of the acinous glands and Bowman's glands, with inflammation and fibroid degeneration of the mucous membrane, resulting in atrophy; (4,) Michel believes that accessory sinus disease, and Bosworth that purulent rhinitis of childhood, stands in causal relation to it.

In my opinion, the disease ozæna is of the nature of a tropho-neurosis, which, like acne vulgaris, is generally connected in some obscure manner with sexual development.

It is often regarded as a final stage of chronic hypertrophic rhinitis. Undoubtedly, this is occasionally true, but in the great majority of cases the disease is primarily atrophic. *Ozæna* is a clinical symptom which is almost invariably associated with atrophic rhinitis, and is either due to some specific micro-organism, *e.g.*, the bacillus fœtidus (Hajek), or is simply analogous to the peculiar odour of the secretion of the axilla or feet.

Though ozæna may be present without atrophy, it is never so persistent and marked as in atrophic rhinitis. From its constituting the most prominent symptom complained of in most cases of atrophic rhinitis, the symptom has given the name to the disease; but we may have, though rarely, atrophic rhinitis without ozæna, and ozæna without atrophy. J. Dunn has recently reported a cured case of ozæna which bears on this point. In his patient ozæna had been present from eighteen to the twenty-fifth year, but had been entirely absent for four years. He found that the atrophic process had extended to the whole of the mucous membrane, and he considers that the disease is essentially a definite inflammatory process of the mucous membrane.

Wyatt Wingrave directs attention to the remarkable atrophy of the lymphoid tissue in the nasal and nasopharyngeal tissues, and in fifty-six out of sixty cases found complete atrophy of the faucial and pharyngeal tonsils. He has also demonstrated that the sticky secretions and crusts are not muco-pus, for the elements of pus are wanting. The secretion is mucus with epithelial cells, multi-nucleated lymphocytes, with some staphylococci and other bacteria.

Though heredity appears to have some influence, and it is often associated with anæmia and the strumous diathesis, ozæna is essentially a disease of puberty and young adult life, and the majority of cases are found in females.

Symptoms.—The patient usually complains of no pain, but of dryness of the nasal passages and naso-pharynx, alternating with profuse discharge when the masses of inspissated secretion come away; but the characteristic and intensely foul odour is not perceived by the patient, who has generally more or less completely lost the sensation of smell. The nose is often flattened and broad, with sunken bridge. Pain at the back of the eyes, or over the bridge of the nose, is sometimes present.

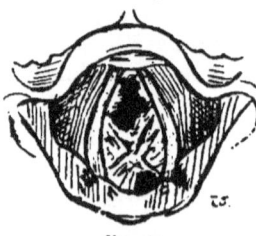

FIG. 85.
Collections of greenish inspissated secretions in *Laryngo-tracheal Ozæna*.

On *inspection*, the most noticeable features are the width of the nasal passages from atrophy of the mucous membrane, so that if the copious sticky secretion, which hangs in strings across the passages or collects in greenish gray masses, is removed, we can often see the pharyngeal wall. Collections of dry secretion are found in the posterior pharyngeal wall, and on examining the larynx and trachea we often find them similarly affected, and that collections of dark green secretion occupy the trachea, the upper surface of the cords, or stretch across the glottis, causing more or less complete hoarseness or aphonia. The view that these are simply collections of inspissated secretion, which have fallen into the glottis from the naso-pharynx, is probably incorrect. When the nose alone is affected, the breath from the mouth will not have the characteristic odour.

Diagnosis.—The bony structures never become necrosed, and the absence of any dead bone or perforation of the bony septum are points which enable one to differentiate between true ozæna and tertiary syphilitic disease, attended with accumulation of *muco-purulent* secretion, and which often closely simulates ozæna in the intense sickly odour which accompanies this necrotic process.

Treatment.—In this, as in all forms of rhinitis, a general tonic and hygienic treatment must be adopted whenever there is any indication of impaired health. But ozæna is very commonly associated with apparently perfect general health.

In the *local* treatment, which at best is usually only palliative, the first essential point is the removal of the crusts of inspissated secretion. This, to be effectual, requires patience and care, but is readily accomplished by projecting a stream of a simple alkaline wash on to the crusts at considerable pressure. The spray should be directed by the eye, with the nasal speculum *in situ*, and with the aid of a good illumination, and should be continued till every particle of secretion has been washed off. Large dry crusts may be gently removed by forceps, or by a probe covered with cotton wool. Similarly, the post nasal douche must be used to get rid of all crusts in the naso-pharynx. The best lotion to use is either simple warm water with a little bicarbonate of soda or sanitas added, or Dobell's solution.

The second point is to consider the best method of increasing the functional activity of the atrophied mucous membrane. In my own practice, I have found that spraying the nasal passages with a 1 per cent. solution of "iodic-hydrarg." gives satisfactory results. It is intensely stimulating, and a hypodermic injection of morphine must be given to allay the pain which lasts an hour or two.

Gottstein's method of plugging the nares with medicated cotton-wool is tedious but fairly effectual, while Greville MacDonald has devised an ingenious plan of inducing increased vascularity of the mucous membrane by respiration through the nares partially occluded by a piece of rubber tube containing a loose cotton wool plug. Freudenthal claims excellent results from internal massage. Bryson Delavan states that the secretions are rendered more abundant by the application of the negative pole of the galvanic current to the affected parts. I have also used with advantage the galvano-cautery as originally recommended by Fränkel, and endorsed by Morell Mackenzie and Sajous. The repeated application of trichloracetic acid to the turbinated bodies will often result in great improvement.

Fig. 86.
Nasal Douche.

Thirdly, the patient must be directed to use an alkaline nasal douche daily to prevent the accumulation of the sticky mucus and the formation of crusts. Warm water at 90° F., containing 1 or 2 per cent. of bicarbonate of soda and common salt, answers admirably for cleansing.

A simple hand-ball douche answers well, but the patient must be directed to use no force, to inject up the most blocked nostril, allowing the fluid to escape by the other side, and to be careful to have the lotion at the proper temperature. Without great care there is always a risk of setting up *otitis media*, and for this reason the douche should only be ordered for cases like ozæna, and should be followed by a soothing, oily antiseptic spray, or, as Christopher Heath recommends, a snuff composed of iodol and borax (1 to 7).

FIBRINOUS OR CROUPOUS RHINITIS.
MEMBRANOUS RHINITIS.

Symptoms.—Fibrinous or croupous rhinitis is attended with the formation of a tough membrane on the nasal mucosa. It is probably due to the action of some form of micrococcus, and the membrane is analogous to the so-called diphtheritic membrane of scarlet fever. The attack may be ushered in with chilliness and rise of temperature, or there may not be any general symptoms at all, as in one case I observed in a boy. There is free mucous discharge from the nose, which becomes more or less completely blocked by the tough membrane.

The affection tends to be protracted, and the membrane reforms several times after removal.

Burn Murdoch has reported a case in which the attacks recurred six times, at intervals varying from one month to a year, each attack lasting about a week or a fortnight. The first attack began in November with symptoms of an ordinary cold. In a few days the nose was completely blocked, and there was a copious muco-gelatinous secretion and numerous fibrinous casts were shed, the nose and face being much swollen and painful, but, as in my case, there was no rise of temperature. The subsequent attacks varied in severity. There were no paralytic sequelæ. W. F. Robertson's examination of the membranous casts of this unique case showed that they were composed mainly of fibrin containing numerous round cells. There were some epithelial cells suggesting shedding of the whole depth of mucous membrane. Sections stained by Löffler's method showed no micro-organisms, and Gram's method only revealed a few groups of micrococci.

As usual there were no paralytic sequelæ nor anything

in the symptoms suggestive of diphtheria, but Dundas Grant states that one of his cases was followed by post-diphtheritic symptoms, and others were infected.

Diagnosis from diphtheria would be very difficult at the outset, though, of course, the absence of the pronounced constitutional depression of nasal diphtheria would soon clear up any doubts on the point.

Treatment consists in treating the general symptoms, and locally in the removal of the false membrane by forceps, and applying simple astringent and antiseptic aqueous solutions.

RHINITIS ŒDEMATOSA.

Rhinitis œdematosa is described by Mulhall, who has had six cases to which he applies this term. It consists of a serous infiltration into the connective tissue of the inferior and middle turbinated bones, which is sometimes migratory, and it appears to be due to irregularities of digestion in neurotic subjects. It is not an inflammatory affection and would be more correctly termed *coryza œdematosa.*

The only treatment advised is local scarification and attention to the digestive tract. Any abnormal conditions in the nose should also be attended to.

RHINITIS CASEOSA.

Rhinitis caseosa *(coryza caseosa)* originally described by Duplay, consists in a blocking of the upper regions of the nasal fossæ with caseous matter similar to that found in some sebaceous cysts. The cheesy matter may form in considerable quantity, even causing facial deformity, and may lead to anosmia. Lennox Browne states that in his experience it is usually associated with some caries of the ethmoid bone.

Chapter XIX.

HYPERTROPHY OF THE PHARYNGEAL TONSIL.

AND NASO-PHARYNGITIS.

HYPERTROPHY OF THE PHARYNGEAL TONSIL.

POST-NASAL ADENOIDS.

THE collection of lymph follicles on the roof and posterior wall of the naso-pharynx, known as the pharyngeal or Luschka's tonsil, generally participates in the hypertrophy and thickening of the mucous membrane in the naso-pharyngeal catarrh which is always associated with chronic rhinitis. But, in many cases, the adenoid hypertrophy is considerable, and the "post-nasal growths" give rise to special and characteristic symptoms, the clinical importance of which, first emphasised by W. Meyer, is now widely recognised by practitioners.

Etiology.—Chronic rhinitis and all conditions which lead to hypertrophy of the faucial tonsils are undoubtedly predisposing causes, and thus hypertrophy of the faucial and pharyngeal tonsils generally go together, while both usually undergo atrophy in adult life. Thus post-nasal adenoids are almost confined to young children, and although they sometimes persist or develop after the age of twenty they usually occur between the ages of five and ten, especially in males. Heredity has some influence in their development, and in some instances they are probably congenital.

Measles, scarlatina, the gouty and rheumatic diathesis, nasal obstruction, and all conditions which predispose to catarrhal rhinitis, are stated to be important factors in their production.

Symptoms.—In children the *facial appearance* is in itself almost pathognomonic. The bridge of the nose becomes broad and flattened, while the alæ nasi are indrawn, the naso-labial fold disappears, and the inner canthus being drawn down, the upper eyelids droop; and thus with open mouth and buccal respiration the little patient presents an expression of vacuity and dulness which is further increased by the inability to fix the attention (*aprosexia* of Guye) and the defective hearing, which are almost constantly marked features.

Speech is nasal and toneless, while stuttering and stammering are sometimes due to post-nasal growths.

Deafness.—The ears should always be examined. From extension of the catarrh, or by direct pressure of exuberant vegetations, the orifices of the Eustachian

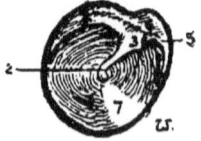

FIG. 87.

The normal right membrana tympani. (1,) The membrana; (2,) The handle of the malleus, ending in the umbo; (3,) short process; (4,) the posterior and (5,) the anterior fold; (6,) membrana flaccida; (7,) the bright spot.

FIG. 88.

The appearance presented by a depressed (right) membrana tympani. The handle of the malleus is foreshortened and the short process very prominent. The posterior fold is abnormally prominent. The bright spot is more diffuse and less distinct towards the periphery of the membrane.

tubes become blocked, and then from the gradual absorption of the air in the middle ear the tympanic membranes often become very much depressed, or from *otitis media purulenta*, perforation of the drumhead may result.

Respiration is peculiarly noisy and snuffling, and, though the nasal passages are seldom completly blocked, is chiefly conducted through the mouth—buccal respiration. Breathing is particularly noisy at night, and suffocative night-terrors often occur. A peculiar barking cough, chorea, or even convulsions, are among the neuroses that may be directly excited by adenoid growths, whilst defective growth and all the evils of mechanical obstruction to respiration in the young should be borne in mind.

Soft adenoids bleed very readily, and thus considerable quantities of blood may be coughed up, or pass into the stomach and be vomited.

Examination may be made either by posterior rhinoscopy, or by palpation with the finger.

If the posterior nares can be inspected, we find either a group of pinkish gray gelatinous looking masses, or one large mass with irregular surface, growing from the roof and posterior wall of the naso-pharynx. With small growths, only a part of the post-nasal space is blocked up, but the amount of growth varies greatly. Very frequently the orifices of the Eustachian tubes and the upper part of the post-nasal apertures are concealed by their presence (see *Plate V., Fig. 3*).

Palpation alone has often to be relied on for the purpose of diagnosis in children, the right fore-finger, protected either by a finger-guard, or by a cork or napkin held between the teeth, being quickly passed up behind the soft palate and swept over the whole post-nasal space so as to diagnose the presence, extent and situation of any adenoids. In children adenoids are gelatinous, soft, and pliable, and readily bleed, but they often have a firm fibrous base, while in adults the hypertrophy of the pharyngeal tonsil is deeper red in colour, firmer in texture, and less exuberant.

Eustachian synechiæ are attributed by Robertson to adhesions, with subsequent contraction, between the Eustachian orifices and the pharyngeal tonsil.

Prognosis.—Patients almost invariably show a remarkable improvement within a very short time after adequate treatment, both as regards general health and mental condition. Percy Jakins reports a remarkable instance of this, that of a lad 5 feet 3 inches in height at seventeen years of age, who increased to 5 feet 10 inches, and proportionately in weight, within two years after removal of enlarged adenoids and tonsils.

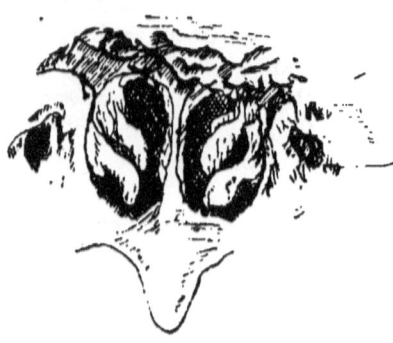

FIG. 89.
Rhinoscopic Image showing distortion of the left Eustachian orifice from the cicatricial contraction of synechiæ.

Treatment.—Before referring to the treatment of postnasal adenoids, it may be well to direct attention to the extreme importance of dealing with all the associated conditions, or those which stand in causal relation to the naso-pharyngeal growth, firstly, the general health, secondly, the local conditions present, such as rhinitis and various causes of mechanical obstruction in the nose, and thirdly, we often have to deal with mischief set up in the middle ear.

Non-operative Measures.—It is not always necessary or desirable to operate in cases which do not present marked symptoms calling for immediate relief. Besides attention to general health, the local use of an oil atomiser containing eucalyptol, terebene, or other antiseptic in solution is useful. If the secretion of muco-pus is abundant, the nasal passages and post-nasal space should be frequently cleansed with a dilute alkaline wash.

The use of astringents, such as chloride of zinc (grs. x ad ʒj), applied by a cotton-wool holder, will sometimes prove highly beneficial.

FIG. 90.
Lennox Browne's Finger Guard and Curette.

FIG. 91.
Dalby's Scraper.

Operative treatment for the removal of growths is necessary in the large majority of cases, and, while almost devoid of danger, is most satisfactory in its results when skilfully and effectually performed. In children, at any rate, a general anæsthetic is desirable. The choice of the anæsthetic for these operations is at present an open question. At the Central London Throat Hospital nitrous-oxide gas is generally relied on, but it involves very rapid operating. By nitrous-

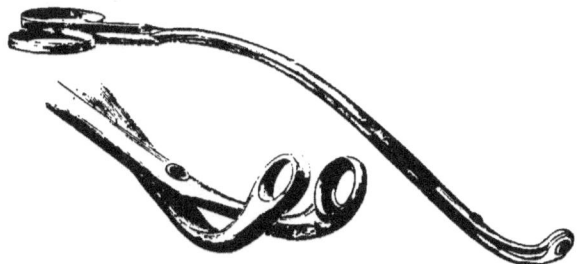
FIG. 92.
Mark Hovell's modification of Löwenberg's Post-Nasal Forceps.

oxide gas with ether we obtain a longer anæsthesia. Chloroform is generally preferable, given till anæsthesia is just complete, care being taken to avoid pushing it to the abolition of the laryngeal reflexes. The patient should be in the *recumbent position*, with the head hanging well down over the end of the table, so that blood may escape by the nose and not enter the larynx. In older patients the use of cocaine does away with the necessity for a general anæsthetic.

The growths may be removed either by scraping with the finger nail, or with Dalby's steel finger nail, or by means of Löwenberg's forceps, but it is generally better to use a curette. Gottstein's is the most useful for removing the large masses in the vault and posterior wall, and has the advantage of being a very safe instrument. It is passed well up behind the soft palate, and, as it is drawn down, is made to include and cut through the hypertrophied tonsil.

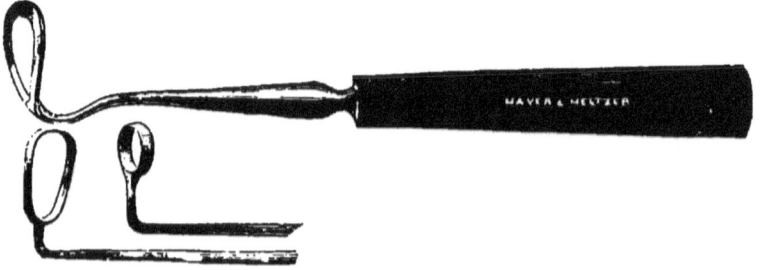

FIGS. 93, 94, 95.
Gottstein's, Hartmann's and Bronner's Pharyngeal Curettes.

Bronner's curette may be used with advantage for removing adenoids in the fossæ of Rosenmüller, it being safer and less likely to injure the Eustachian orifices than Charters Symonds', or Hartmann's, although in skilled hands there is little risk of such an accident. After using the curette, the finger should be introduced to complete the operation, and any adenoids on the Eustachian orifices are best removed by the finger alone. Secondary hæmorrhage seldom occurs, but one fatal case has been reported. It may be necessary to plug the naso-pharynx. For tough growths it is often necessary to use Löwenberg's forceps, or one of the many modified forms of this instrument. For the smaller hypertrophies of adenoid tissue in adults, which are attended with general thickening of the naso-pharyngeal mucosa, nothing is so satisfactory as the galvano-cautery, a guarded naso-pharyngeal electrode, such as Bosworth's or Macdonald's being used. The patient must be kept indoors for a day or two after the operation, and the parts cleansed daily with a coarse alkaline spray, either the "lotio alkalina" or a 1 to 2 per cent. solution of sodium chloride, or bicarbonate.

Naso-pharyngitis, till recently regarded as a common affection, is usually a complication simply of rhinitis and pharyngitis, and dependent on an extension of the morbid process observed in these diseases. Owing to the accumulation here of muco-pus from the nasal passages, it often leads to the supposition that it is this part of the upper respiratory tract which is principally affected. This is only true when hypertrophy of the pharyngeal tonsil is present.

Chapter XX.

DISEASES OF THE SEPTUM.

PERFORATION, DEFLECTIONS AND SPURS, ETC., EPISTAXIS, ABSCESS.

PERFORATION OF THE SEPTUM.

Perforation of the septum was found by Zuckerkandl eight times in 150 autopsies, and is, therefore, more frequent than is generally supposed. In many cases it causes no symptoms and is without significance, and certainly it cannot be held that perforation of the cartilaginous septum is a proof of syphilis. It may be due to simple ulceration in dry rhinitis, from picking with the finger nail, to idiopathic periostitis, or result from an abscess following a blow. Perforation sometimes occurs in lupus and malignant disease, and then the cause is obvious. But the commonest cause is syphilis. When only the cartilage is involved, symptoms are absent, but if the perforation involves the bony septum, fœtor is almost invariably present in very marked degree. The bridge of the nose only becomes sunken when the nasal bones are diseased. I have known the whole septum to have completely disappeared without any falling in of the nose at all.

SEPTAL DEFLECTIONS, CRESTS, AND SPURS.

Deviation of the septum may almost be regarded as its normal condition, for it is only in comparatively few persons that it is absolutely straight. Morell Mackenzie, in an examination of 2,152 skulls in the Royal College of Surgeons, found a deviated septum in 1,657, or about 75 per cent. In about 40 per cent. of the cases the deviation was to the right, in 30 per cent. to the left, while the remainder were sigmoid or zigzag. In some cases, however, the deviation causes such marked symptoms, either from simple mechanical obstruction and its consequences, or by setting up nasal neuroses, that the condition calls for treatment.

Etiology and Pathology.—As regards the etiology of these deformities of the septum, there is much difference of opinion. Morgagni and others believe that the deviation is frequently congenital, but Zuckerkandl,

FIG. 95.
Septal deflection (2,) with nasal obstruction, due to bony prominence (*), in association with turbinal enlargement.

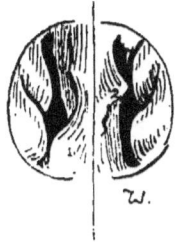

FIG. 97.
The same after cocaine has caused shrinking of the turbinals, the bony prominence (1) and the septal deflection (2) remaining unaltered.

in his elaborate investigations, found none under seven years of age. Ziem attributes the deviation to falls or injuries, causing irregular and defective growth of the other bones, while Lennox Browne, MacDonald, and Zuckerkandl agree in attributing to blows and injuries the chief responsibility for the condition, and my own observations have led to the same conclusion.

The deflection, in patients presenting symptoms, is generally most marked in the triangular cartilage, especially in its anterior portion. It may be a simple twist of the anterior end to the right or the left, or it may be a sigmoid deflection, obstructing alternately the right and left passages.

As the result of perichondritis following fracture of the cartilage or its dislocation from its attachment to the vomer, a cartilaginous ridge may be formed antero-posteriorly or obliquely. A deflection on one side generally has a corresponding depression on the other.

In other cases we find simple kinks, spurs, or rounded prominences projecting into the passage, and approaching or coming into contact with the lower turbinal.

Symptoms.—As a consequence of the partial nasal obstruction, inspiration causes a rarefaction of the air behind the obstruction, with consequent over-filling of the vessels, and constant hyperæmia which results in chronic rhinitis.

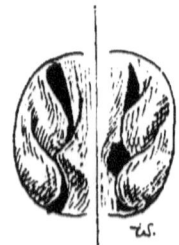

FIG. 98.
C. T.; Septal deflection, associated with hypertrophic rhinitis and paroxysmal sneezing, which disappeared after treatment.

The other nasal passage, even if normally patent, is seldom equal to moistening the whole of the inspired air, and, therefore, is liable to inspissation. Thus, in a patient recently sent to me by a medical practitioner, the chief complaint was pain and dryness of the left nasal passage, for which he had been recommended to artificially close this side, especially on going to bed. The whole trouble was due to a deviation of the septum occluding the right passage, and, therefore, the condition of the patient was, of course, only aggravated by closing both sides.

I have usually found that in addition to the symptoms caused by rhinitis and post nasal catarrh, the patient is very liable to repeated attacks of bronchitis and laryngitis. As a consequence of rhinitis and the extreme irritability of the nasal mucosa set up by a septal spur, patients sometimes suffer from paroxysmal sneezing.

Diagnosis.—This should be obvious on examination of the nasal passages. The only conditions that are liable to cause confusion are syphilitic perichondritis, and

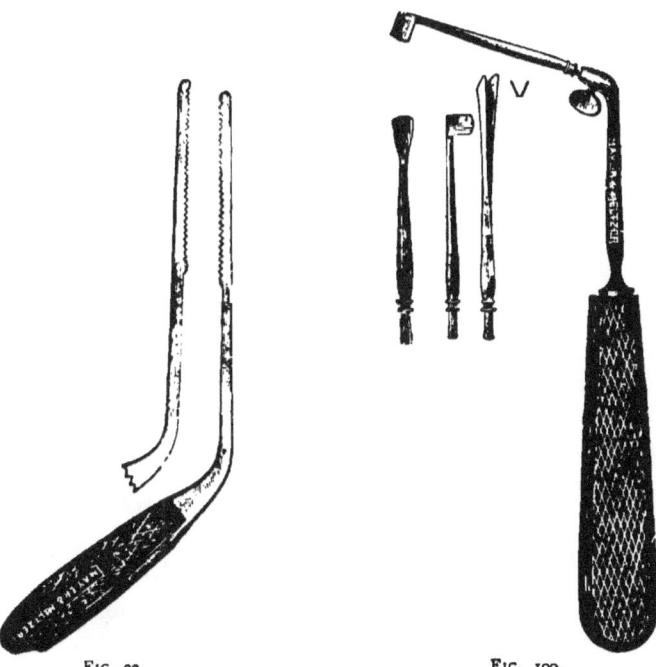

FIG. 99.
Bosworth's Saw.

FIG. 100.
MacDonald's Gouges for Antero-posterior and vertical ridges.

hypertrophy of the vascular tissue of the septum. The former is inflammatory, the latter is soft and fluctuating and is reduced by the application of cocaine.

Treatment.—In cases where there is only a narrow ridge, the simplest method of relieving the condition consists in reducing it by means of the galvano-cautery.

In a great many patients the obstructive symptoms are due to an associated hypertrophy of the inferior turbinated body, and sometimes the treatment of the turbinal hypertrophy (see *page* 195) is all that is needed.

FIG. 101.
Curtis' Trephines for septum.

But it is often necessary to remove the ridge or spur by mechanical means. We may then use with advantage MacDonald's gouges, or Bosworth's nasal saw. Seiler uses burrs, and Curtis trephines, driven by a small electro-motor, or by a hydraulic motor as used by dentists.

MacDonald's method is as follows: "A single linear incision is made over the most prominent point of the neoplasm, well down to the cartilage. With a raspatory the perichondrium, with its inseparable mucous membrane, is then turned up and down sufficiently to expose the portion to be removed. Next, the superabundant cartilage is separated with a gouge, or saw if it prove to be

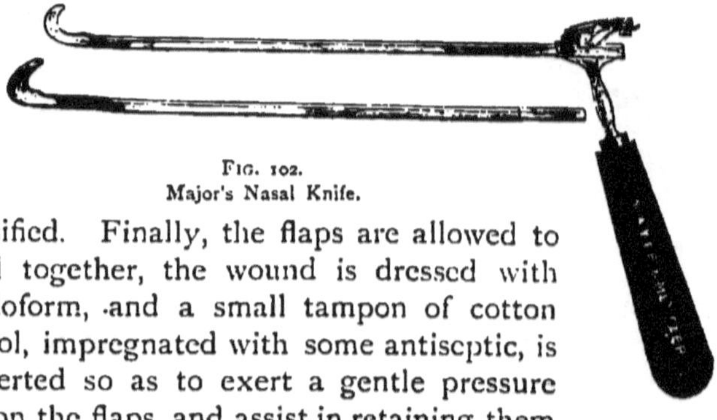

FIG. 102.
Major's Nasal Knife.

ossified. Finally, the flaps are allowed to fall together, the wound is dressed with iodoform, and a small tampon of cotton wool, impregnated with some antiseptic, is inserted so as to exert a gentle pressure upon the flaps, and assist in retaining them in their position. Healing frequently takes place by first intention" (Diseases of the Nose).

Favourite instruments are Carmalt Jones' spoke-shave, and Major's nasal knife, which are placed in position behind the spur, and then firmly drawn forward, cutting through the obstructing cartilage. These knives have been specially designed for that class of case in which a long crest or spur is found running in a horizontal direction from before backwards along the lower third of the septum narium near the floor of the nasal chamber. There are numerous other instruments, introduced by different rhinologists, but they are mostly modifications of one or the other of the forms already alluded to. Adams' or Hewetson's septum forceps, or Hill's dilator.

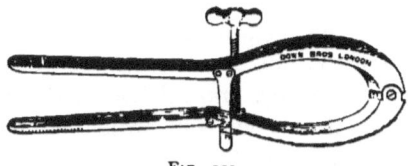

FIG. 103.
Hill's Dilator.

should only be used when it is desired to correct a bony deformity, or when a deviation causes external deformity, for they can hardly be expected to relieve obstruction so effectually as the methods of removing the redundant or obstructing tissue.

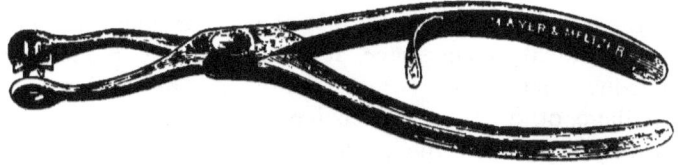

FIG. 104.
Jarvis-Steel Septum Forceps.

Punches, such as Steele's, are liable to cause sloughing and perforation.

Brysan Delavan first elevates the mucous membrane

and periosteum, and then if the soft tissues are replaced a perforation is recovered from. J. N. Roe has modified Steele's forceps, so that by means of a regulator on the handle the punch just stops short of perforation of the mucous membrane.

Chiari for simple deviations resects the septal cartilage in the form of a triangular flap, and keeps the flap in an over-corrected position by means of iodoform plugs for two or three weeks. Moure, of Bordeaux, strongly advocates electrolysis, and in the discussion on his paper at the Rome congress Rosenfeld supported him in this practice, stating that "cartilage melts away under its action like butter under a hot sun. Only one operation is needed, and this is applicable to bony growths as well, if the bone is not too dense to be penetrated by a needle" (*Jour. Laryng. and Rhin.*, vol. viii., No. 5). Scissors are useful only for well projecting small spurs.

As a rule cocaine is the only anæsthetic required in the operation, and as hæmorrhage is generally rather free it presents special advantages in tending to check it, and in that by its use we avoid the difficulties that such hæmorrhage causes when the operation is done under general anæsthesia.

If forceps are used, or if the portion to be trephined is considerable, a general anæsthetic will be necessary, as the operation is painful. When a considerable surface removal is necessary, it is desirable to reflect the mucous membrane before removing the underlying cartilage, otherwise an extensive cicatrix is formed, and causes dryness of the passage and discomfort to the patient.

EPISTAXIS, BLOOD-TUMOURS, AND ABSCESS.

HÆMORRHAGE from the nose may be due to (1,) *injuries* from blows, etc.; (2,) *local affections* of the nasal passages, hyperæmia, ulceration in malignant disease, or from foreign bodies; (3,) *systemic affections,* such as anæmia, purpura, scurvy, Bright's disease, portal congestion, cirrhosis of the liver, enteric fever, measles, scarlatina, diphtheria, pneumonia, etc.; or (4,) it may be vicarious and occur at the menstrual periods only.

The hæmorrhagic spot is most frequently at the anterior inferior part of the septum. The bleeding point may be at any part of the nasal fossæ, or there may be general oozing. If the bleeding is from the anterior part of the passages, the blood escapes from the anterior nares; but when it is more deep in origin it passes into the naso-pharynx, and may be swallowed and subsequently vomited in large quantities, or it may pass into the glottis and be coughed up and thus simulate hæmatemesis or hæmoptysis.

In some cases dependent on general conditions, *e.g.,* plethora, portal congestion, or renal disease, the loss of blood is beneficial within limits, but when it is persistent or profuse it is necessary to check it.

Treatment.—In epistaxis, if the simpler domestic methods have failed, the bleeding point should be sought for. If in the anterior part of the septum, simple compression of the alæ with the finger and thumb may suffice to stop it, or the bleeding point may be touched with the galvano-cautery at a black heat.

For a less defined or a more deeply-seated source of hæmorrhage, cold or iced salt and water douches may be tried, or a spray of hazeline. When other means have failed the nose should be plugged. For this purpose the nicest thing is Cooper Rose's inflating plug; Bellocq's sound is very convenient, or a soft rubber catheter can be passed through the nasal passages till it appears in the pharynx, when it is seized with forceps, drawn into the mouth, and the post-nasal plug of absorbent wool or lint (about the size of a walnut) is tied on. The plug is then drawn into place till it has occluded the posterior orifice of the nasal fossa on the corresponding side, and the plugging is completed from the front in the usual manner.

BLOOD-TUMOUR AND ABSCESS.

Following injuries to the nose we sometimes find, on either side of the septum, a dark red or purplish swelling of fluid blood effused beneath the mucous membrane. In course of time the blood may undergo absorption or organisation, or it may degenerate, forming an abscess, which may become very chronic.

Blood-tumour if small may become absorbed, but if large the blood should be evacuated and the nose kept aseptic by suitable sprays or lotions. Abscess of the septum should of course be evacuated, and appropriate after-treatment adopted.

Chapter XXI.

DISEASES OF THE ACCESSORY SINUSES.

THE ANTRUM OF HIGHMORE, THE ETHMOIDAL CELLS, THE FRONTAL SINUS, THE SPHENOIDAL SINUS.

DISEASES OF THE ANTRUM OF HIGHMORE.

Acute Inflammation.

Acute inflammation of the antrum of Highmore is exceedingly rare. Semon has reported his own experiences in an attack following influenza, and refers to several similar cases, all due to influenza, observed by others. With the exception of one case that had to be opened, they all recovered spontaneously.

The **Symptoms** noted were: (1,) a sense of intolerable distension in the cheek, with redness, tenderness, and swelling over the region of the antrum; (2,) sudden and violent increase in the pain on blowing the nose, coughing, or sneezing, the pain being limited to the affected region; (3,) the exudation in the antrum increases till the distension causes sudden evacuation of the turbid, greenish, sero-purulent fluid through the ostium.

Chronic Empyema.

Chronic empyema of the antrum was described by John Hunter, but the credit of recognising the frequent occurrence of the affection without the classical signs of pain, swelling, and tenderness is due to Ziem, who published his first paper in 1886.

Etiology and Pathology.—The maxillary sinus is lined by mucous membrane in continuity with that of the nose through the ostium maxillare, while it is in close relation with the teeth, the roots of the first and second molars forming rounded prominences in the floor of the antrum. Thus disease of the antrum may be due, on the one hand, to diseased conditions of the nasal passages, and, on the other, to caries of the teeth, the relative importance of these two sources of the affection being at present a moot point.

We may recognise as causes of antral empyema :—

(1,) simple rhinitis, extending by continuity to the antrum, with resulting stenosis of the ostium. It is especially prone to occur in this manner in syphilitic, chronic hypertrophic, and atrophic, rhinitis, when these are attended with muco-purulent discharge, and in scarlatina, measles, erysipelas, influenza, and mercurial ptyalism.

(2,) Occlusion of the ostium by mucous polypi (Bayer) or deviated septum pressing on the middle turbinal. We must bear in mind that polypi and a polypoid condition of the mucous membrane around the ostium is often the result rather than the cause of antral empyema.

(3,) Pus may descend from the frontal or the ethmoidal sinuses, and escape into the antrum. It is probable that many obscure cases of antral disease are caused by simple catarrhal secretions descending by the infundibulum, and failing to find a ready exit through the normal orifice, pass into the antral cavity and there undergo decomposition from retention.

(4,) The majority of cases are secondary to caries of the teeth, the sockets of which project into the antral cavity.

Symptoms.—We divide cases of antral empyema into two classes: (1,) those in which there is more or less free exit to the pus through the ostium maxillare; (2,) those in which this natural opening is obstructed.

In the first class, patients generally complain of a chronic cold in the nose, with constant or periodic discharge of considerable quantities of fœtid pus. Not unfrequently their attention is first attracted by the blocking of the nose, either from the swelling of the mucosa, or from the presence of polypi. The symptoms

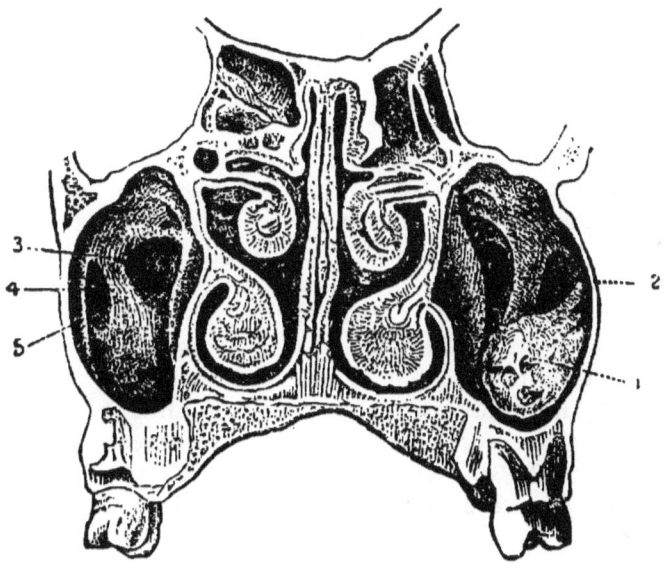

FIG. 105.
Transverse section through the nasal cavities and maxillary sinuses. The floor of the antrum is at a lower level than that of the nasal passages. A polypoid growth (1), and bony trabeculæ (2), and other irregularities (3, 4, and 5) are seen in the cavities of the antra (Zuckerkandl).

are generally unilateral, and unilateral blennorrhœa is very suggestive of antral empyema. Pain and tenderness over the region of the antrum, or dental neuralgia is occasionally noticed. The unilateral discharge of pus is especially liable to occur on rising from bed, or on stooping, or on lying with the head on the sound side.

When the pus is permanently retained, the antral cavity may become distended, and we observe a smooth, hard, tender swelling on the cheek corresponding to the antrum; but distension of the antrum is generally due to either cystic disease or myxomatous polypi.

Diagnosis.—The mucous membrane of the middle turbinal and around the *ostium* is red and swollen. The most constant sign however, is the presence of a small bead of pus appearing beneath the anterior extremity of the middle turbinal. If after wiping the pus away it re-appears in the spot immediately, the evidence of antral empyema is conclusive, and the pus may often be made to appear by gentle pressure over the antrum, or by the patient tilting his head to the sound side. The pus is generally extremely fœtid, in which case the odour is perceived by the patient, except when the antral disease is a complication of ozæna—a very frequent association according to Robertson.

The condition of the teeth will sometimes throw some light on the case. They may be all quite sound except those corresponding to the diseased antrum, which may be carious or have "dropped out."

There are three other methods by which the diagnosis may be rendered more certain :—

(1,) By *transillumination*, as suggested by Voltolini and developed by Heryng. A small four-candle-power electric lamp is placed in the mouth which is closed, the room having been rendered absolutely dark. In a normal patient the nasal passages and the cheeks become illuminated by a rosy-red transmitted light, the patient often perceiving a sense of light himself. Any difference in the amount of light transmitted is estimated, and if pus be present in the antrum, or it is the seat of a solid tumour, there is generally an obvious difference, more especially in the infra-orbital region, which on the the affected side is in shadow, and the patient himself will often perceive light on the sound side only. But any inflammatory thickening of the mucous membrane of the antrum, or a difference in the thickening of the bony walls may produce an umbra on the affected side, while a cystic tumour may rather increase than decrease the brilliancy of the transmitted light.

(2,) *Exploratory puncture* by means of a Lichtwitz trocar and cannula, or with a strong exploratory hypo-

Fig. 106.
Lichtwitz' Trocar.

dermic needle passed through the wall of the inferior meatus; or by drilling through the socket of a decayed tooth after its extraction.

(3,) Irrigation through the natural orifice. Moreau Brown, having applied cocaine to the mucous membrane of the middle meatus, introduces a syringe with a curved nozzle through the ostium maxillare, and irrigates the antral cavity with a solution of peroxide of hydrogen, (1 in 12). If pus is present it is driven out and fills the nose as a white foam.

Treatment.—The treatment of cases of antral disease requires much care and perseverance on the part of both patient and attendant. In all methods the object is the same, viz., to evacuate the pus and cleanse the alveolar mucous membrane with some antiseptic and stimulating application.

There are five alternative methods of procedure:—

(*a*,) Treatment through the natural opening is advocated by a few rhinologists. A special syringe with a curved nozzle is directed through the ostium and the solutions injected. Moreau Brown irrigates in this manner for diagnostic purposes only. If treatment is attempted through this orifice it should be enlarged and the anterior portion of the middle turbinal may have to be removed so as to allow freer access. Garel has obtained good results by daily irrigation of boric acid solution through the natural opening into the antrum.

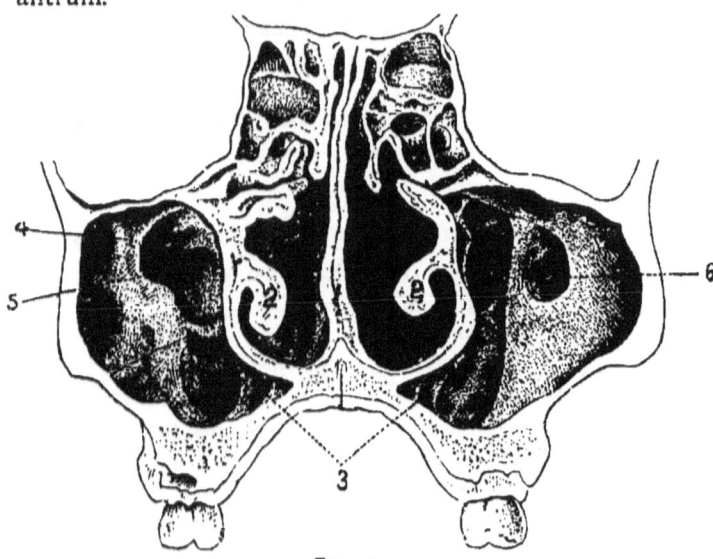

FIG. 107.
Transverse section showing irregular development of the maxillary sinuses (Zuckerkandl.)

(*b*,) Treatment by irrigation through an artificial orifice in the anterior portions of the outer wall in the inferior meatus. This is not generally to be commended since it is impossible to secure good drainage, the floor of the autrum being well below the floor of the nasal passage. Moreover, the orifice is necessarily too small to allow of currettage.

(*c*,) Removal of a tooth and opening into the antrum by drilling the socket. This is frequently done, but though the position is a good one for drainage food is liable to enter the cavity. It is therefore necessary to have a dental plate made to fit the patient, with a spur which projects into the cavity and keeps the artificial opening from closing till the diseased cavity has been cured.

(*d*,) The mucous membrane having been reflected a hole about the size of a sixpence may be trephined into the antrum in the canine fossa, passing directly backwards with a slightly upward direction, or the opening may be made with a chisel and mallet. This is the method advocated by Robertson.

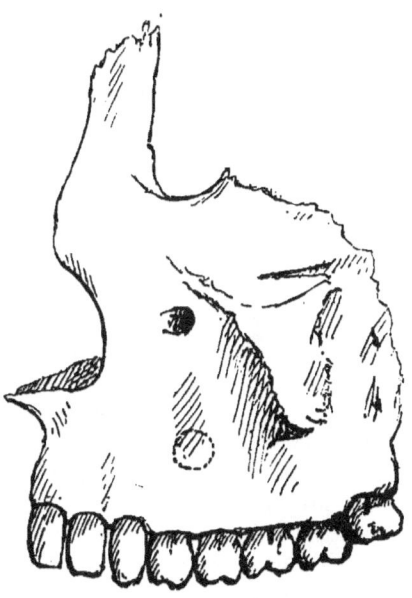

Fig. 108.

To show the place where the superior maxilla should be trephined for antral disease.

The antral cavity should be thoroughly washed out with warm water, and when all pus and inspissated secretions have been removed, it must be washed out finally with some non-irritating antiseptic, *e.g.*, listerine. The syringing should be carried out three times daily at first, and then twice a day till the discharge has completely ceased. After a time we may use some stronger and more stimulating injection such as chloride of zinc, (gr. v ad ʒj) or a weak aqueous solution of iodine, and the cavity gradually allowed to close.

By introducing the finger or a probe through the opening, any bony partitions which cause retention of pus and prevent drainage, may be broken down. Robertson uses a small electric search-light to explore the condition of the antrum.

(*e*,) The method introduced by Scanes Spicer is a modification of Robertson's. The opening into the antrum is made as in Robertson's method. The inner surface of the antrum is then curetted so as to remove any granulation tissue, etc., after which with the finger introduced into the cavity to act as a guard, Krause's trocar and cannula are passed down the inferior meatus of the same side, and one or two large perforations into the antrum are made well behind the nasal duct (see *Frontispiece*). The antrum and nose are then irrigated with warm boracic lotion, and the cavity of the antrum tightly packed with creolin gauze for forty-eight hours. At the end of forty-eight hours the gauze is removed and no drain is used at all, but free irrigation is used three times a day and the patient instructed to blow air through the antrum, first by the nose and then by the mouth, and to use boracic lotion frequently in the same way. By this means the antrum is more effectually cleansed and a source of irritation in the drainage tube is dispensed with.

Dr. James Swain has obtained excellent "through drainage" by passing a tube up the tooth socket and bringing it out through the ostium at the anterior nares.

Cystic and other Diseases.

Various tumours are found in the antrum, viz., myxomatous polypi, malignant new growths, dentigerous cysts, and cystic degeneration of the lining membrane. They generally give rise to an obvious external swelling, and require removal.

DISEASES OF THE ETHMOIDAL CELLS.

Etiology.—These cells are liable to distension from occlusion of their openings, with the formation of a mucocele or empyema. The latter disease is usually the result of chronic rhinitis, erysipelas, or due to the escape of pus from the antrum through the opening in the neighbourhood of the ostium, generally in association with nasal polypi.

The **Symptoms** are, pains in the infra-orbital region and in the nose, with displacement of the eye outwards and upwards or downwards if distension has occurred. Pus escapes either anteriorly from beneath the middle turbinal, or posteriorly into the naso-pharynx. There is considerable danger of extension to the cribriform plate with resulting meningeal abscess.

The **Diagnosis** can often be made only by excluding antral disease, which, however, is frequently coincident.

Charters Symonds has observed that in addition to the frequent presence of polypi and fungating granulations, the middle turbinal is itself found to be soft and gelatinous on removal with punch forceps, a valuable point in the differential diagnosis from frontal sinus empyema in which this condition is not observable.

Treatment.—In some cases the cells can be curetted through the nose and irrigated with mild antiseptics, but when distension and displacement of the orbit has occurred it may be necessary to gain access to the diseased bone through the inner corner of the orbit.

DISEASES OF THE FRONTAL SINUS.

A catarrhal condition of the lining membrane of the frontal sinus is usually associated with coryza, and is then the cause of the discomfort so often experienced in this region.

The sinus may be invaded by certain insects, or it may be the seat of mucocele or empyema.

The causes of frontal-sinus empyema are obscure; it is sometimes due to influenza, erysipelas, measles, or scarlatina, blows on the nose, or from blocking of the infundibulum by polypi.

Symptoms of Frontal Sinus Empyema.—If the infundibular opening is patent, there is *continuous* discharge of pus from the opening, just in front of the ostium maxillare. The chief symptoms are supra-orbital neuralgic pain, or pain at the root of the nose, and discharge of pus from the nose.

Richardson Cross considers that "where a continuous discharge from the nose is complicated with frontal headache, and where occasional cessation of the discharge is associated with increase in the headache, and particularly with swellings or evidence of inflammation over any surface of the frontal sinus, empyema of this cavity is certain and operation for its relief should not be delayed."

When the infundibulum is blocked, the symptoms are greatly aggravated, and in course of time swelling and redness and tenderness are found over the root of the nose and inner angle of the orbit, while the eye may be pushed downward and outwards. From distension of the sinus the brain may be compressed, and meningitis is liable to occur.

Treatment consists in maintaining the patency of the infundibulum, if possible, and irrigating the sinus through this opening. Lichtwitz reports a case cured after a year's treatment in this manner.

When the infundibulum is occluded it is necessary to make an opening over the sinus in the forehead, treating it on the principles laid down for maxillary sinus empyema, and to restore the passage to the nose.

DISEASES OF THE SPHENOIDAL SINUSES.

These may be the seat of empyema, and, as in the other accessory sinuses, the pus may either gain exit or be retained.

Symptoms.—If the pus flows, indefinite, deep-seated pain, with post-nasal discharge, is present. When the sinuses become distended the pain is greatly increased, and symptoms of basilar meningitis are liable to supervene. From pressure on the optic nerves, sudden amaurosis may occur; then ophthalmoscopic examination will generally reveal choked discs. The eye is sometimes pushed forward.

Treatment.—The sphenoidal sinuses may be opened by a sharp curette passed backwards and upwards above the middle turbinal. It has also been suggested that the opening should be made through the mouth into the floor of the sinus forming the roof of the naso-pharynx just behind the choanæ (see *Frontispiece*). Of course these operations should only be attempted by an expert. Indeed, the symptoms of abscess of the sphenoidal sinus are so difficult and obscure that only one experienced in rhinology would be able to come to any definite diagnosis.

Dundas Grant in the discussion on empyema of the nasal accessory sinuses at the meeting of the British Medical Association at Bristol, 1894, maintained that a discharge of pus along the upper part of a nasal passage between the middle turbinal and the septum, should always lead to the suspicion of sphenoidal sinus disease, and advised the use of Lichtwitz' long cannula for the purpose of irrigating the sinus after it had been punctured with a trocar.

Chapter XXII.

CHRONIC INFECTIVE DISEASES.

SYPHILIS; TUBERCULOSIS; LUPUS; GLANDERS; RHINOSCLEROMA.

SYPHILIS OF THE NOSE.

Inherited Syphilis.

The *early form* occurs within the first three months of life, and assumes the form of a catarrh with tumefactions of the nasal mucosa with consequent "snuffles." The discharge of mucus or muco-pus is irritating to the anterior nares, producing excoriations and fissures, and tending to form crusts in the nasal passages which become somewhat fœtid. At this period necrosis of the cartilage is rare.

The *late form* manifests itself between the age of five and puberty. It corresponds to the tertiary period in the acquired form, and is, therefore, characterised by gummy infiltration, caries, and necrosis of the cartilage of the septum, the vomer, and the turbinated bones, with fœtid discharge and with consequent deformities.

Acquired Syphilis.

Primary sore from inoculation with the finger nails is very rare. The chancre does not differ in aspect from chancre elsewhere, but the secretion from the resulting catarrh is apt to collect and become inspissated.

Secondary syphilis gives rise to slight symptoms, chiefly nasal *catarrh* with tumefaction of the Schneiderian membrane. *Mucous patches* may occur on the septum and inferior turbinated bodies.

Tertiary syphilis may assume the form of a localised gumma of the septum or turbinated bodies, a firm, circumscribed, red swelling. More generally we find extensive ulceration and suppuration, with caries or necrosis of the cartilaginous septum, the vomer, and the turbinated bodies. The discharge is considerable, purulent, bloody, yellowish-green, but is apt to collect and form fœtid greenish-black crusts. If bony caries has occurred, the stench is indescribable and most penetrating to those around, though to the patient the sense of smell is diminished or altogether lost: a probe will detect diseased bone concealed by the greenish necrotic tissue.

The disease is generally bilateral, though often more advanced on one side. Pieces of necrosed bone may be separated from time to time with the discharge, and in consequence of the contraction of the subcutaneous connective tissue, or from the partial or complete destruction of the nasal bones and septum, the nose becomes characteristically broad and sunken, "saddle-backed," or the whole of the cartilages and tissues of the alæ may be lost. Following gummy infiltration, the bone may undergo partial absorption without necrosis (Schech).

Diagnosis.—The diagnosis should rarely present much difficulty. Syphilis may attack any part of the nose, but the loss of tissue is generally confined to the septum. The history and concomitant lesions will generally be enough to confirm the diagnosis which the nasal features suggest.

Syphilis in the nose must be differentiated from lupus, tubercle, malignant disease, and ozæna (see p. 255).

Treatment.—In addition to the general treatment appropriate to the particular phase of syphilis present, local treatment is generally necessary.

In young children at the breast, syphilitic catarrh by interfering with nasal respiration may render suckling impossible. The child should be fed by the spoon till the condition has yielded to the internal administration of mercury in small doses. If, as is often the case, the infant declines all nourishment from any but the natural source, we may spray a weak solution (2 per cent.) of cocaine, or a solution of menthol (20 per cent.) in olive oil or liquid vaseline into the nares before putting the child to the breast, as recommended by McBride.

The special treatment in syphilis of the nose consists in the use of alkaline and cleansing douches to keep the passages free from the accumulations of secretion. In tertiary ulceration insufflations of aristol or iodol tend to keep the fœtor under, and calomel fumigations may be useful in checking the progress of the disease; but with necrosis of the bone it is useless to expect any means of overcoming the stench to succeed. The dead bone should be gently removed if possible.

TUBERCULOSIS OF THE NOSE.

Etiology and Pathology.—Tubercular disease in the nasal passages is extremely rare, and, probably without exception, is always secondary to tubercular disease in the lungs or larynx.

Weichselbaum in 146 autopsies on patients who had died with tubercular disease, found only two cases in which the nose was implicated. I have only seen one case myself, but it is probable that a systematic examination of the nose in cases of advanced phthisis would show that it was less rare than is supposed. M. Herzog, who has reported 10 cases of his own, and has collected and reviewed the literature of all the cases recorded, 80 in number, finds it occurs in the form of (1,) neoplasms; or (2,) ulcers; or (3,) a combination of both forms.

(1.) *Tubercular deposits* are found from the size of a poppy seed to a large walnut, although they are seldom larger than a split-pea. They are of irregular outline, rounded or elliptical, of reddish or grayish-yellow colour, surrounded by elevated soft margins in which sometimes miliary tubercles are seen. They are soft and friable, and bleed easily.

(2,) *Tubercular ulcers* here, as in the pharynx, are shallow, irregularly round or oval, with soft slightly elevated margin, with a grayish-yellow base filled with caseous tubercles. They are usually situated on the nasal septum, near the anterior border. The tubercular deposit, of course, consists of granulation tissue, with giant cells, and contains tubercle bacilli.

Symptoms.—At first the symptoms are very indefinite, but as the deposit increases nasal obstruction may occur. Very soon the small miliary tubercles ulcerate, and then a muco-purulent discharge comes on, often bloodstained, and after continuing a short time becoming very fœtid. In fact cases of tubercular disease of the nose closely resemble ozæna, from which Hajek distinguished them by finding the characteristic bacillus. The absence of pain is remarkable.

Diagnosis.—Tubercle may be differentiated from tertiary syphilis by the presence of tubercle in other parts, by the progressive character of the lesion in spite of anti-syphilitic treatment, and by the presence of the bacillus, and from its selecting the cartilaginous septum and not the osseous. It must also be distinguished from lupus, ozæna, malignant disease, and chronic glanders.

Treatment.—The local treatment consists in thorough removal of the deposits by scraping, and the subsequent application of chromic acid, lactic acid, or the galvano-cautery, followed by some antiseptic insufflation daily, such as aristol, europhen, etc. The disease is extremely prone to recur after a short period. Of course, the usual general treatment for tuberculosis should be pursued at the same time.

LUPUS OF THE NOSE.

Etiology and Pathology.— Intra-nasal lupus is extremely rare, unless we include those cases of lupus of the alæ nasi which invade the vestibular portion.

The nodules occur generally as multiple, small, hard, elastic tubercles, covered with dry brown scales of inspissated secretion, or greenish-yellow pus. Beneath the scab will be found the characteristic ulcers, round or oval, cup-like, with raised indurated margin, tending to cicatrise at one part and extend in another, and are painful if pricked with a probe. They are generally situated on the septum, near the anterior inferior extremity, or close to the floor, and sometimes on the anterior part of the inferior or middle turbinals. When the crusts separate there is a certain amount of bleeding, followed by sero-purulent discharge which may be very offensive if the crusts have been long retained.

The bones are very rarely, if ever, involved in the necrotic process, but the cartilaginous septum generally becomes perforated sooner or later.

Symptoms are mainly those of nasal obstruction, with a limited amount of intermittent discharge. Pain is generally absent or only slight. Other symptoms will be present, of course, if the pharynx and larynx are involved.

Diagnosis is generally easy, inasmuch as intra-nasal lupus is usually associated with lupus of the skin or pharynx.

It is distinguished (1,) from syphilis by the destructive process not extending to the bony structures, the tendency to heal, and the slight amount of discharge, and the absence of that intensely sickening odour of syphilis, which is not removed by the most careful cleansing; (2,) from malignant disease by its slow progress and tendency to heal, and from epithelioma, though not from sarcoma, by its occurrence in the young; (3,) from nasal tuberculosis by the appearance of the deposit, and its slow course.

For other points in the differential diagnosis, see Lupus of the Pharynx and Larynx (p. 61).

Treatment consists in removal of the crusts, and thorough cleansing of the passages, followed by curettement and the application of lactic acid and other local agents, as in lupus of the pharynx. General treatment by tonics, cod liver oil etc., must of course be combined with local treatment.

GLANDERS.

Etiology and Pathology.—Glanders is a contagious disease of horses and cattle, due to the specific micro-organism the *b. mallei*, but is occasionally met with in men who generally contract it from infected horses, and thus it is usually seen in ostlers and grooms. Morell Mackenzie records a fatal case, which resulted from a diseased horse sneezing when being driven in a hansom, some of the secretion coming into the face of the patient who was inside.

It occurs in the acute and chronic forms. The disease may be chronic at the outset, subsequeutly passing into the acute variety.

Chronic glanders closely resembles tertiary syphilis of the nose, but it is extremely rare. The mucous membrane of the nasal fossæ is slightly swollen and may be painful, and is covered with dirty scabs. There is a peculiarly viscid, offensive, muco-purulent discharge, and when ulceration supervenes it is serous. The disease may extend to the pharynx, back of the tongue, and larynx.

A large proportion of cases end fatally in six to eight months. As the bones are often implicated in the ulcerative process, the diagnosis from syphilis is rendered all the more difficult, and often depends on the absence of any improvement from anti-syphilitic treatment.

Acute glanders is the commoner form of the disease in man, but, according to Böllinger, who collected 120 cases, it is less frequently localised in the nose, than is the case in the horse.

Symptoms.—The general symptoms are ushered in by a general febrile condition, with headaches, rigors, and pains of a rheumatic character in the limbs. In a day or two a pustular nodular infiltration occurs in the nasal mucosa, with a profuse glairy discharge. The nodules ulcerate, and the discharge becomes more viscid and muco-purulent, while the nose externally becomes red, painful, and swollen.

A papular eruption resembling small-pox, comes out on the face and limbs; diarrhœa, profuse sweating, vomiting, and general prostration, usually end in coma and death in less than three weeks.

Diagnosis.—The general symptoms closely resemble acute rheumatism or typhoid fever at the onset, but the extremely adynamic condition should serve to exclude acute rheumatism and tertiary syphilis, while the absence of other features serve to eliminate typhoid fever and small-pox. From anthrax it would be distinguished by the absence of the "charbon pustule."

Treatment is practically hopeless, and must be conducted on general principles, for which the reader will consult works on general medicine.

Locally, Elliotson reported success in stopping the

nasal discharge, by injecting a solution of 2 grains of creasote in a pint of water three times a day. Other mild antiseptic agents will suggest themselves.

It is hardly necessary to emphasise the extreme importance of the most rigid prophylactic measures to prevent the inoculation of healthy individuals.

RHINOSCLEROMA.

Etiology and Pathology.—The cause of this rare affection, originally described by Hebra, is unknown, though Stepanow and Cornil have described bacilli which they claim as the specific infecting agent. Lemeke holds that rhinoscleroma is identical with Störk's blennorrhœa.

It generally commences on the upper lip near the alæ nasi in the form of small, hard, raised nodules, which very gradually spread to the nasal passages. The nodules are hard, tender on pressure, covered by normal skin, and not inflammatory. It may begin in this manner in the nose, or may take the form, *ab initio*, of diffuse thickening of the mucosa without showing any nodules. The mucous membrane is normal or slightly reddened in colour, smooth and shining, and very hard. The infiltration consists of small round cells which subsequently become spindle-shaped. There is no inflammation, no discharge or pain, and no ulceration.

Symptoms are entirely local, viz., stiffness and nasal obstruction, and the course of this disease is very chronic. It may spread to the pharynx, or even to the larynx, in which case respiration may become embarrassed.

It generally occurs in advanced adult life, but Semon's case was in a boy of 14, a Guatemalan. Very few cases are recorded as having occurred in England, but it seems fairly prevalent in Austria and in some parts of Egypt, South America, and India.

Diagnosis must be made from syphilis and malignant

disease, in the one case by the fact that the affection is not altered by syphilitic remedies, and in the other by the local character and its slow development.

Treatment.—The mechanical obstruction may be reduced by the galvano-cautery or knife. No treatment hitherto introduced has appeared to be of the slightest use in arresting the disease, till Stoukovenkow, of Kiew, had successful results in one case with interstitial injections of Fowler's solution of arsenic (1 to 12 per cent.). The treatment was continued for fifteen months, and no less than 222 injections were made.

Chapter XXIII.

NEOPLASMS OF THE NOSE AND NASO-PHARYNX.

MUCOUS POLYPUS, AND BENIGN NEOPLASMS; MALIGNANT NEOPLASMS.

MUCOUS POLYPUS.

THE only common form of benign neoplasm of the nasal passages is the mucous polypus. Zuckerkandl found polypi in as many as one out of every eight or nine autopsies in which the nasal passages were carefully examined, from which it must be inferred that they exist in a great many people without producing definite symptoms.

Etiology and Pathology.—The cause of these polypi is very obscure. Morell Mackenzie refuted the suggestion that they are the result of chronic catarrhal conditions, pointing out that polypus is rare before the age of sixteen, while chronic nasal catarrh is especially common in young children, but Lennox Browne remarks that though simple catarrh is an affection of childhood, hypertrophic rhinitis is almost confined to adult life. They are more frequent in men than women, and though rare before puberty, they are met with even as early as the seventh or eighth year.

Doubtless there is a causal connection between suppuration in either the maxillary or ethmoidal sinuses and mucous polypi. The ethmoid bone is usually diseased at the seat of the polypi; indeed, Woakes believes that polypus is only a symptom of bone disease, which he terms "necrosing ethmoiditis." But there is seldom any clinical or pathological evidence of *necrosed* bone, what we find is an osteitis, rarefying and formative, as shown in the drawing (*Fig.* 109) of a section from the base of a mucous polypus. Zuckerkandl in reference to this question states that he has not observed necrosis of the bone in a single case, but that he has seen the bony part which exists in the base of some polypi become elongated and softened. The layers of the periosteum undergo inflammatory changes which result in this *osteitis* and in hyperplasia of the fibrous connective tissue, by œdematous infiltration of which the ordinary polypus is formed. Thus we may often detect rough or bare bone, or a sense of softened carious bone conveyed by the contact of a probe with the bony trabeculæ at the base of the polypi. By the irritant action of muco-purulent secretion, inflammatory and degenerative changes are induced, which not only originate the sodden, water-logged mucous membrane so often associated with mucous polypi, but may cause some alteration in the vascular supply, with consequent exudation and œdema such as is believed by Hopmann to be the prime factor in their causation.

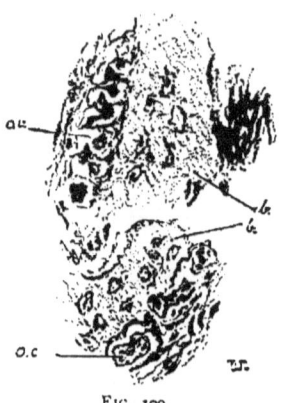

FIG. 109.
Section from the base of a mucous polypus, showing a bony spicule, *b*; with osteoblasts, *ob*; and osteoclasts, *oc*.

Thus the so-called myxoma or mucous polypus is composed of fine meshes of areolar tissue filled with fluid containing serum-albumen and mucin, and covered with the ciliated epithelium of the mucous membrane when small, though this normal ciliated epithelium is generally lost and replaced by stratified epithelium as the polypus becomes larger. They contain a variable amount of glandular tissue, generally only a few mucous glands being present, yet rarely the glandular element may be so preponderant as to constitute a fibro-adenoma.

From cystic degeneration of the mucous glands we get one form of cystic polyp, but in other polypi, in which the glandular element is almost or completely absent, simple liquefaction of the contents may give rise to one large cyst. Thus we may differentiate the varieties of mucous polypi by the terms :—

| Fibromacœdematosa | Fibro-adenomacœdematosa |
| Cysto-fibromacœdematosa | Cysto-fibro-adenomacœdematosa |

Though from cases coming under treatment, we find that mucous polypi nearly always arise from the middle turbinal, or from the margin of the *hiatus semilunaris*, Zuckerkandl found from post-mortem examinations that they are very frequently situated in the superior meatus, or on the superior turbinal.

Symptoms.—The most prominent symptom is stuffiness or obstruction in the nose, varying, of course, with the size of the polypi. The patient can often feel something which "flops" to and fro on inspiration and expiration, and may be distinctly conscious of some loose body in the nasal passages.

The nasal secretions are in excess, and the constant sniffing and running at the nose are extremely annoying. If nasal respiration is much obstructed, mouth breathing with its attendant evils is present, the voice becomes nasal, muffled and without resonance, and the sense of smell is lost. As the growth increases in size it may press upon and obstruct the orifices of the accessory sinuses (with consequent retention of the secretions of the antrum, frontal sinus) or of the nasal duct, giving rise to lachrymation and epiphora. In this manner we may sometimes find empyema of the antrum when frontal sinus empyema has occurred. I have only once met with frontal sinus empyema which could be attributed to blocking of the infundibulum by a polypus, and certainly polypus is more often the result than the cause of antral empyema.

Patients generally find the symptoms aggravated in damp weather, partly owing to the hygroscopic properties of the polypi, and partly to the effect of damp on the rhinitis which almost invariably co-exists.

Catarrhal deafness often supervenes from associated naso-pharyngeal catarrh, or from direct pressure on the Eustachian tubes, and from these catarrhal conditions and obstruction in the lymphatics between the dura mater and the mucous membrane of the nasal fossæ, the dizziness, loss of memory, and inability to fix the attention (*aprosexia* Guye), arise.

Mucous polypi, if large, often cause deviation of the septum, but while the nose may appear broadened from some œdematous infiltration, external bony displacement is hardly ever seen.

True nasal neuroses are not uncommonly set up. Thus Hack enumerates nightmare, persistent cough, hemicrania, and epilepsy, as especially liable to occur when the anterior inferior portion of the lower turbinal is pressed on, while, as in Voltolini's classical case, asthma is sometimes due to the presence of a polypus.

Objective examination seldom leaves any room for doubt as to the diagnosis. If the polypus has attained any size, and is situated anteriorly, it will be seen as a characteristic semi-translucent, smooth, grayish, gelatinous body, occupying the middle meatus, or reaching down to the inferior meatus, or even presenting at the anterior nasal orifice. Unless very large it is generally possible to determine its seat of origin, which, as stated above, is generally beneath or from the free border of the middle turbinal. Less frequently its attachment is higher up, but never does it originate from the inferior turbinal, and hardly ever from the septum.

When the polypus is large only one may be seen, but they are generally multiple, and, more often than not, bilateral. Mucous polypi are sometimes very large. Thus Zaufal reports the removal of one more than 6 inches in length, and weighing over $3\tfrac{1}{2}$ ounces. In many cases no large polypus is present but there are numerous small polypi the size of currants, or smaller, beneath or attached to the free edge of the turbinal. Sometimes the free border of the middle turbinal has undergone polypoid degeneration when no definite pedunculated polypus is present.

When a polypus is deeply seated it may often be made to come to the front by the patient blowing the nose vigorously. But when thus growing far back, or when the polypus has extended backwards towards the posterior nares, the diagnosis can only be made by a posterior rhinoscopic examination, which shows the polypus protruding from the choanæ into the naso-pharynx, or even completely filling it, so that it hangs down and appears below the soft palate. There should be no difficulty in distinguishing a polypus in this situation from (1,) hypertrophy of the inferior turbinal; (2,) symmetrical adenoid

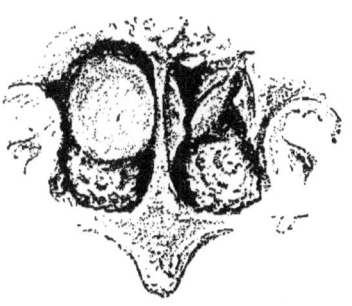

FIG. 110.

Posterior rhinoscopic view, showing moriform hypertrophy of the inferior turbinals, partly concealed on the (patient's) right side by a smooth mucous polypus. Hypertrophic tissue is likewise seen on the vomer.

growth on the septum; (3,) post-nasal adenoids; but it is not easy to make a diagnosis by inspection alone from those rare cases of fibrous polypi growing from the nasal cavities towards the naso-pharynx.

Large polypi, which have undergone compression by surrounding structures, tend to lose their usual œdematous appearance, and become more solid and fibrous on the surface, and, if coming well into the anterior nares, may be red and lobulated, and simulate papillomata in appearance.

Examination of these polypi with a probe will always show them to be soft and freely moveable, and that they are not connected with the septum or inferior turbinal, while they are readily penetrated by a sharp needle.

Diagnosis.—Mucous polypi must be differentiated from fibrous polypi, cancer, and sarcoma, all of which are painful, bleed freely when probed, are firm in texture, and, if large, produce bony displacement. Chronic abscess of the septum and bloody tumour, which results from blows, are generally bilateral, and are situated on the septum. Papilloma, hypertrophy of the lower turbinated body and foreign bodies, will hardly give rise to confusion on careful examination, while cartilaginous and osseous tumours are hard and present other distinguishing characteristics. Christopher Heath states that in his experience one of the commonest errors in diagnosis is that of taking the convex side of a deviated septum for a growth.

Treatment consists (1,) in removal of the polypi, and (2,) treatment of the abnormal conditions which have caused their growth.

The only methods of removal which merit discussion are (*a*,) avulsion, and (*b*,) snaring.

Avulsion, by seizing the polypi with suitable forceps, has been extensively condemned by many specialists, who advocate the use of the snare as the only justifiable means. To blindly introduce forceps and tear away whatever happens to have come within their grasp is indeed a disastrous procedure, but when the position of the polypus has been made out, and forceps are used with skill and care, they are very efficient means of getting rid of the growths. There are two conditions in which they should be resorted to, viz., when it is impossible to introduce the loop of a snare owing to the size or situation of the polypus, or when a number of small sessile polypi occupy the middle meatus and can-

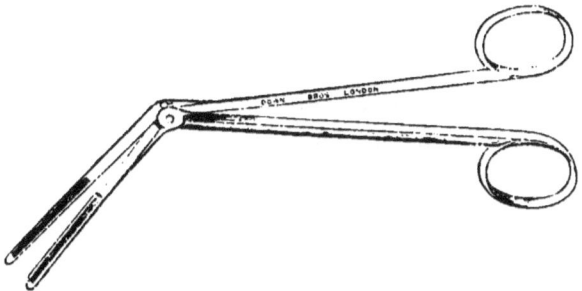

FIG. 111.
Morell Mackenzie's nasal polypus Punch Forceps.

not be snared. These small polypi should be removed one by one by small forceps. For larger polypi, Mackenzie's punch forceps or broad-bladed, serrated forceps with a catch, are the most suitable. The polypus should be seized as near its attachment as possible, and the forceps then twisted round and round till the growth is detached. In addition to the danger of tearing away a portion of the turbinated bones, the disadvantages of the forceps are the pain that their use involves and the hæmorrhage. Bleeding generally ceases in the course of a few minutes, and may be checked by spraying with cold water. If hæmorrhage continues to be free, it may be necessary to plug the nasal passages. (See Epistaxis).

Of course, before polypi are removed, cocaine (10 or 20 per cent.) should be used. This should be applied, as far as possible, to the root of the polypus and to the

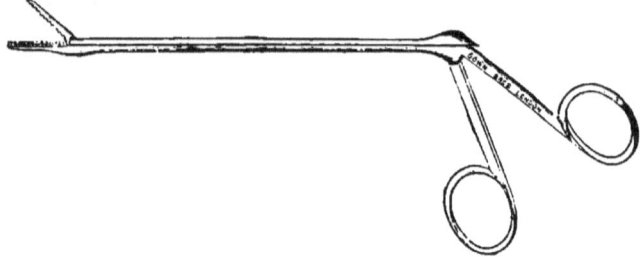

FIG. 112.
Alligator Forceps for removal of small sessile polypi.

mucous membrane of the nasal fossa, and not simply sprayed on to the polypus itself which is practically devoid of sensation. For this purpose a spray with a fine nozzle that can be introduced between the polypus and the inner and outer wall of the nares should be used. In young children it is sometimes necessary to use a general anæsthetic. The posterior nares should then be

FIG. 113.
Nasal Polypus Forceps.

plugged, if possible, to prevent the escape of blood into the naso-pharynx with the necessity of constant mopping.

Snaring is the method of removal that should always be adopted, if possible, as it is far less painful, more under the control of the operator, and involves less hæmorrhage.

The most generally useful form of snare is Mark Hovell's modification of Morell Mackenzie's ratchet snare, as the loop can be tightened slowly and can always be drawn out readily and re-introduced without the trouble of changing the wire. Bosworth's snare is very convenient, and, being slender, can be introduced into very narrow passages. For rapid work no form is more convenient than Krause's snare.

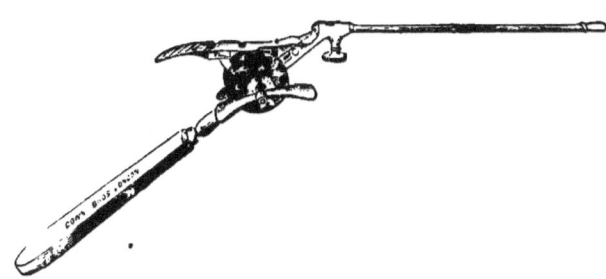

FIG. 114.
Mark Hovell's modification of Morell Mackenzie's Polypus Snare.

A No. 5 to 10 steel piano-wire is generally used for the cold snare. After the application of cocaine, the nostrils being dilated by a nasal speculum, the loop of the snare, which must, of course, be large enough to pass over the whole polypus, should be introduced vertically, the upper part of the loop being insinuated between the outer wall of the nasal passage and the growth. Then, turning the snare to the horizontal, it should be passed up to the root of the polypus by gentle movements, and gradually tightened as the narrow pedicle is reached. The loop having thus been placed round the pedicle as near the base as possible, it should be tightened till the pedicle has been cut through, when the growth can be removed.

Polypi which are situated far back or have passed into the naso-pharynx, often prove most difficult to snare. If it can be brought forward by the patient blowing the nose, it may be seized with a hook or small forceps, and held while the snare is passed over the hook or forceps, and thus made to encircle the pedicle. It is sometimes possible to pass the snare along the floor of the nasal fossa to the naso-pharynx, and then, with the aid of the forefinger in the mouth, to get the snare round the dependent growth. In some cases the forceps passed well back can be made to seize the pedicle, and the polypus be got rid of in this way.

The galvano-cautery snare may be used instead of the cold wire snare, the loop being placed in position in the same manner as for the ordinary snare, and the current passed through the wire as it is being tightened. It is said to be followed by less hæmorrhage, and to have a

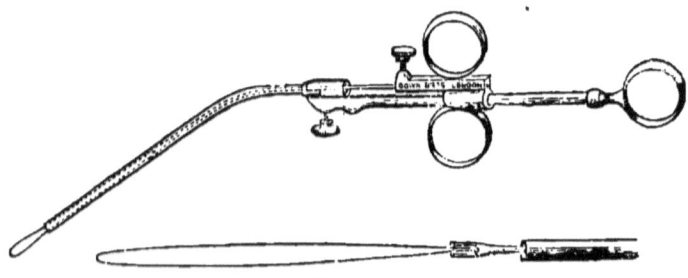

FIG. 115.
Krause's Polypus Snare.

greater tendency to prevent recurrence. But the cold wire snare can be employed without involving much bleeding, provided it is used slowly and skilfully, while it is always necessary to cauterize the stump, even if the cautery snare has been used. Moreover, the cautery snare is generally followed by considerable inflammatory reaction from the burning and scalding of the contiguous mucous membrane as it is burning through the pedicle For these reasons it has largely fallen into disuse, except for those polypi with thick firm pedicles that the cold snare will not cut through. When the passages are very narrow it is occasionally necessary to remove the anterior inferior portion of the middle turbinate by scissors or snare, as recommended by Casselberry.

To prevent the recurrence of the growth it is always advisable within a day or two to deeply cauterise the pedicle with the galvano-cautery, or to apply chromic acid by means of a probe. The mucous membrane around the pedicle is generally in a more or less polypoid condition, and should be cauterised as well. Indeed the small currant-like polypi which sometimes surround the opening of the *hiatus* are often best treated by the application of the cautery alone. The parts may be sprayed with 25 per cent. alcohol as recommended by A. G. Miller, and the patient instructed to employ the spray daily for some time. The patient should also be directed to use an oil atomiser with some antiseptic in solution, *e.g.*, eucalyptol or terebene, grs. x to xx to the ounce of liquid vaseline.

Hypertrophic rhinitis, septal deviations, or other abnormal conditions which may be present must be treated if recurrence is to be prevented. But it is essential that the patient be kept under observation for some time, as even when all growths have apparently been removed at the time of the operation others will often become obvious in the course of a day or two, after the larger mass, which has compressed or displaced them, has been removed.

The general treatment is of considerable importance, as in all cases of rhinitis, and a tonic treatment and change of air will go far to prevent recurrence, and complete the cure.

The other benign neoplasms which are occasionally met with are :—

FIBROMA, an exceedingly rare growth in the nose, but fairly common in the naso-pharynx. They are firm, compact, painful, bottle-shaped growths, covered with smooth light pink or red mucous membrane, and are generally attached to the periosteum of the vomer, the body of the sphenoid, or to the septum. As they bleed readily, in addition to the usual symptoms of nasal polypus they are attended with epistaxis. They may be removed by the snare simply if small, otherwise W. J. Walsham, who has had considerable experience in dealing with these growths, advises splitting

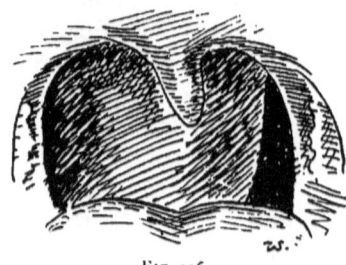

Fig. 116.
Large Fibroma growing from the naso-pharynx.

the soft palate, and, if necessary, partial removal of the hard to give room for the snare. The reader may refer to Prof. Annandale's remarks on the removal of these growths in the naso-pharynx, *Lancet*, 1894, vol. i. p. 398.

PAPILLOMATA always grow from the lower turbinal or the septum, vary in size from a pea to a filbert, and are often multiple. They are pink or red, soft, lobulated, mulberry-like growths which are moveable and bleed readily. They are papillomatous in structure, but are sometimes very vascular, and have then been described as angiomata. I have only met with two cases; one was single, the other multiple and bilateral, and in the latter case there was also (*Fig.* 117, 1) a smooth red vascular tumour the size of half a spanish nut occupying the septum opposite the middle turbinal, similar to those reported by Verneuil, Schadewald, and Flateau. Papillomata

Fig. 117.
Papilloma of septum, 1, 2.

should be removed by the snare, or by the application of chromic acid. Fletcher-Ingals has found the application of thuja occidentalis reduces the growths and tends to prevent their recurrence, which is very liable to occur.

EXOSTOSES are uncommon, and rarely attain any size. They grow from the floor, or from the lower part of the septum, when they simulate deviated septum. They are hard and painless, but may cause obstruction.

OSTEOMATA are sometimes found, but I have never met with a case. There are two varieties, the hard ivory and the soft or cancellous, the former being the most frequent. They grow from the periosteum of the septum or turbinated bones, and are moveable, being attached by a short firm pedicle. The surface is covered with red uneven mucous membrane, and by growth in various directions they may cause displacement of bony structures. They may often be removed by forceps or snare, but sometimes come away spontaneously from atrophy of the pedicle.

ECCHONDROSES are very rare, occurring generally in young males on the septum, and cause obstruction, catarrhal symptoms, and, if large enough, external deformity.

If not too large they should be removed by the cold snare or the galvano-caustic snare.

CYSTS.—Besides the various forms of cystic polypi already referred to, we may rarely meet with cystic growths from the turbinal bodies, and air-containing cavities in the middle turbinal bone have been described by McBride. Hydatids and dermoid cysts have likewise been met with in the nasal passages.

MALIGNANT NEOPLASMS.

Sarcoma is rarely primary in the nose, while *carcinoma* is extremely rare. A sarcoma sometimes takes the form.

of a mucous polypus, which it may so closely simulate as only to be differentiated by microscopical examination. I have known a patient, in whom ordinary mucous polypi had been repeatedly recurring for some years, develop large sarcomatous polypi which eventually involved the whole of the nasal and contiguous structures, an instance of the occasional sarcomatous degeneration of fibromata.

Symptoms.—Primary malignant growths generally develop on the septum. Sarcoma occurs as a single, sessile, soft, lupus-like deposit, with smooth or rugose surface of pink or brown colour, rapid in growth, and very vascular. Cancer, epithelial or encephaloid, commences as a small wart-like growth of dark red or purple colour, and generally ulcerates early.

Malignant growths are attended with fœtid discharge, pain, epistaxis, and grow rapidly. In the early stages, with small defined growth, radical removal would offer some chance of extirpating the growth.

TABLE FOR DIFFERENTIAL DIAGNOSIS.

Chronic Hypertrophic Rhinitis.	Atrophic Rhinitis with Ozæna.	Syphilis.
Age:— Adults, especially males.	*Age:—* Puberty and young adults, especially females.	*Secondary:*—occurs with manifestations of syphilis in other parts. *Tertiary:—*
Symptoms:— Nasal catarrh and obstruction, discharge muco-purulent, and naso-pharyngitis. No pain. Hypertrophy of the inferior and middle turbinals, especially posteriorly, where they form spawn-like masses in the choanæ. No ulcer, no hæmorrhage.	*Symptoms:—* Loss of smell and intermittent discharge of intensely fœtid mucus, and nasal obstruction by dry crusts; both smell and obstruction removed by cleansing. Olfactory fissure wide, bridge of the nose often depressed. Atrophy of turbinal bodies. No ulcer, no hæmorrhage. No destruction of bone.	Gumma, nasal obstruction persistent, no pain, little discharge. Is a large smooth, red, hard, and elastic tumour. It soon ulcerates with muco-purulent discharge. Ulcer round, deep, margins slightly raised, surrounded by areola; floor covered with pus. Not tender to touch. Does not bleed easily. Destruction of bone as well as cartilages. Syphilitic neoplasms are very amenable to treatment by internal remedies.
Tubercle. Generally associated with tubercle elsewhere.	*Lupus.* Generally associated with lupus of skin, in young chiefly.	*Benign Growths.* Polypus freely moveable, gelatinous, translucent, pedunculated growths from the middle or superior turbinal bodies or meatuses. Generally multiple.
Symptoms:— Nasal obstruction absent or slight. No pain. Neoplasms. Ulcers small, ovoid, or irregular outline, margins not raised, surrounded by pale mucous membrane, covered with grayish-white opalescent muco-pus, surrounded by pale mucous membrane. Not tender to touch, and not great tendency to bleed.	*Symptoms:—* Nasal obstruction. No pain, or only slight. Clusters of small, red, firm, elastic growths of slow progress. Ulcers—margin raised, covered with crusts, tends to cicatrise, surrounded by normal mucous membrane. Bleed easily on removing crusts, and are painful when touched. Caries of soft structures only.	*Symptoms:—* Nasal obstruction or discharge. No pain. *Fibroid:—* A single, firm polypus or sessile growth from septum or inferior turbinal. Painful. *Papilloma;—* A cherry red or pink moriform growth on the inferior or turbinal septum, often multiple. Painless nasal obstruction the only symptom.
Malignant Growths. *Age:—* Advanced. Sarcoma at any age. *Symptoms:—* At first nasal obstruction and epistaxis, rather than pain. Later, intermittent.	lancinating pain. Constitutional cachexia. Growths generally unilateral, single, springing from septum, are sessile, red or purple, soft, bleed on touch, not tender. Rapid growth and early	ulceration, with discharge of sanious greenish muco-pus, soon becoming fœtid. Microscopical examination important. Ulcer deep, irregular, covered with muco-pus.

Chapter XXIV.

NASAL NEUROSES.

OLFACTORY NEUROSES, PARÆSTHESIÆ, NASAL COUGH,
VASO-MOTOR RHINITIS, HAY FEVER.

OLFACTORY NEUROSES.

ANOSMIA. — For clinical convenience it is usual to include under the term *anosmia* more or less persistent loss of smell from whatever cause, whether it be due to local conditions of the mucous membrane or to impairment of the nerve structures themselves. For the proper perception of odours it is essential that the odoriferous particles be able to reach the mucous membrane of the upper part of the nasal passages, and that these should be in a moist condition. Therefore any local affections preventing respiration through the nasal passages, or deposits of mucus and secretion on the mucous membrane, or a permanently dry condition of the membrane, will interfere with or completely abrogate the sense of smell.

The terminal filaments of the olfactory nerves may be so altered or impaired, either from a chronic local inflammatory affection, or in atrophic rhinitis, or as the result of injecting very irritating sprays or douches, that anosmia results. It may occur in hysteria, locomotor ataxia, or from basilar meningitis affecting the olfactory bulbs. It is only very rarely that tumours or intracranial hæmorrhages produce anosmia, as it is unlikely that both bulbs would be destroyed. A few cases of unilateral destruction of the olfactory bulb with anosmia are recorded, and in these the left bulb is always the one affected.

Anosmia may result from blows, producing fracture of the cribriform plate.

Increased sensitiveness to smell is sometimes found in hysteria.

PAROSMIA, or the imaginary or subjective perception of odours, is always central, and may occur in hysteria, hypochondriasis, or in epileptics.

SENSORY AND REFLEX NEUROSES.

It may be well to briefly refer here to the physiological reflexes of the nose since, without a proper appreciation of their nature, it is difficult to understand the occurrence of many of the nasal neuroses. The physiological reflexes are sneezing, coughing, lachrymation, and vasomotor changes giving rise to increased secretion. It is well known that bright sunlight often gives rise to sneezing, cough, lachrymation, and rhinorrhœa from excitation of branches of the fifth nerve. Dust and particles of foreign matter and irritating vapours on entering the nasal passages induce like phenomena. Three areas, termed the hyperæsthetic areas, have been determined, situated one on the posterior extremity of the inferior turbinal and the corresponding portion of the septum (J. N. Mackenzie), another at the anterior extremity of the inferior turbinated body (Hack), and the third in the mucous membrane of the vestibule (Sajous), excitation of which is particularly prone to give rise to nasal reflexes; and, to these areas may be added the anterior extremity of the middle turbinated body, the corresponding portion of the septum, and the region of the Eustachian tubes.

These sensitive areas correspond with the ramifications

of the ethmoidal branch of the nasal nerve and the branches from Meckel's ganglion, which are the nerves of sensation to the nasal fossæ.

HYPERÆSTHESIA is generally the immediate local factor in nasal cough, paroxysmal sneezing, hay fever, and hay-asthma, but in all these conditions it is usual to

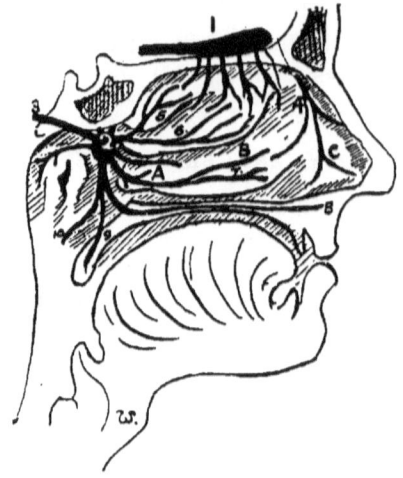

FIG. 118.

Diagrammatic represen'ation of the nerve supply to the interior of the nose. A, The posterior sensitive area (J. N. Mackenzie); B, The middle sensitive area (Hack); C. The anterior sensitive area (Sajous). 1, Olfactory bulb with branches descending to the region of the superior and middle turbinated body; 2, spheno-palatine ganglion; 3, Vidian nerve; 4, external filament of the ethmoidal branch of the nasal nerve; 5 and 6, branches of the spheno-palatine ganglion to the superior and middle turbinated bodies; 7, branch of the anterior palatine to the inferior turbinated body; above it is seen the naso-palatine branch to the septum nasi; 8, anterior palatine nerve; 9, middle; and 10, posterior division of palatine nerve.

find more or less definite abnormal conditions of the nasal mucous membrane associated with hyperæsthesia at one or the other of the sensitive spots in the nasal passages.

ANÆSTHESIA from implication of the fifth nerve may be due to hysteria, cerebral tumours, or intra-cranial syphilis (Schech). Incomplete anæsthesia is not unusual in old standing large polypi.

Nasal Cough.

Nasal cough is usually hard, persistent, ceasing during sleep, and unattended with expectoration, or, with any lung condition to account for the cough. It will often be excited by touching the sensitive parts with a probe, and is analogous to that cough that is often induced by the passage of the Eustachian catheter.

Schadewaldt has directed attention to the interesting fact, that, when the posterior part of the nose or nasopharynx is irritated, the sensation is subjectively referred to the larynx.

I would emphasise the importance of bearing in mind the possibility of a persistent cough, without bronchial or pulmonary disease, being a nasal cough, for it occurs with far greater frequency than is generally suspected, while on the other hand such a cough may be the earliest indication of locomotor ataxia (see p. 175).

Nasal cough should be treated on the same lines as other reflex nasal neuroses (see treatment of paroxysmal sneezing, p. 263).

Vaso-Motor Rhinitis.

Vaso-motor rhinitis, with simple erectile swelling and vascular engorgement, occurs in neurotic persons or in gouty subjects *(see Plate V., Fig. 1)*. It is generally an indication of nervous prostration, or of imperfect digestion. It is sometimes associated with slight abnormalities in the nasal passages.

The fulness of the nasal mucosa may be accompanied by general redness of the nose, or of the face, while in rare instances, the redness and swelling of the nose which comes on in heated rooms or after meals, is due to pressure of the distended middle turbinated bodies on the septum. Of this condition I have only seen one instance, and the periodical redness and swelling of the nose, which rendered the patient almost unfit for society, was completely relieved by cauterising the middle turbinated bodies.

Treatment consists in attention to the general health, while, *if necessary* only, the galvano-cautery may be used, one or two linear cauteries being made over the turbinated bodies or septum.

Paroxysmal Sneezing and Hay Fever.

Etiology and Pathology.—Clinically we may arrange "paroxysmal sneezers" in three classes:—

(1,) Those in whom the lesion is entirely peripheral, such as the earlier and milder forms of true hay-fever, in which the symptoms appear only when the patient is exposed to pollen, etc. For these, local treatment rarely fails.

(2,) Those in whom the symptoms, in the first place induced by peripheral irritation, have become more or less persistent, even in the absence of any pollen or specially irritating particles.

(3,) Idiopathic cases, not due to peripheral irritation at all, and in which the affection is a central neurosis.

Exciting causes.—Paroxysmal sneezing may be excited by any irritation of the fifth nerve, either directly from irritating dust in the nose, or reflexly by the action of bright sunlight.

Hay-fever is simply paroxysmal sneezing set up by particular forms of irritating dust, to wit, pollen grains. Attention was first seriously directed to hay-fever by Bostock in 1819, who endeavoured to prove that the symptoms were due to the solar rays; but the researches of Blackley* have proved that it is due to the action on

---

* Cited by Mackenzie, "Hay Fever."

the mucous membrane of the pollen grains of certain natural orders of plants: Ranunculaceæ, Caryophyllaceæ, Compositæ, Geraniaceæ, Leguminosæ, Graminaceæ, Papaveraceæ, Fumariaceæ, Cruciferæ, etc., but, especially the pollen of Graminaceæ, anthoxanthum odoratum, meadow grass, barley, wheat, and oats. Roses, and in America, Roman wormwood, ragweed, etc., have also a peculiar tendency to excite this affection. Any form of dust impinging on the nasal mucosa in susceptible individuals, may bring on the symptoms.

Predisposing causes.—As the exciting causes of hay-fever are practically universal, and every one must be exposed to them, while comparatively few suffer, it is obvious that individual predisposition is necessary for the symptoms to appear. It is generally found in dwellers in cities, in the educated classes, and especially in those persons of the neurotic temperament, in other words, in the class of persons who are prone to neuroses, and therefore, to some extent it is hereditary. Hay fever is more common in men than in women, and the symptoms are increased in severity in hot weather.

Local Conditions.—The exciting causes are universal, the predisposing are very common, while the affection itself is *relatively* rare. Thus, a third factor is generally necessary for the occurrence of hay fever, and this is found in the abnormalities and morbid conditions of the nasal passages. We are indebted to Roe, of New York, and Daly, of Pittsburgh, for the recognition of these local conditions as causes of hay-fever, viz. :—(1,) hypertrophic rhinitis ; (2,) spurs and bony projections of the turbinals or septum ; (3,) septal deviations ; (4,) polypi and adenoid hypertrophy of the naso-pharynx ; (5,) peculiarly sensitive areas.

Thus, for the occurrence of hay fever, three factors are generally necessary :—

(1,) The predisposing constitutional condition;
(2,) An external irritant;
(3,) A pathological condition of the nasal mucous membrane.

Symptoms.—The symptoms of hay-fever come on in Great Britain about the middle of June, and in America about the middle of August, while simple paroxysmal sneezing may occur at any period. At first only a slight itching of the inner canthus is observed, and watering of the eyes. In a day or two some irritation in the nose with watery discharge comes on, with nasal obstruction, due, in fact, to vaso-motor rhinitis, and pricking and dryness in the throat. Very soon attacks of sneezing supervene, and recur with increasing frequency without the pleasant sense of relief that is usually associated with a good sneeze. The conjunctivæ become injected, the eyes bloodshot, and the nasal passages more or less blocked up by the swelling of the mucous membrane and turgid erectile tissues. The soft palate and uvula become relaxed, and all the painful symptoms of "clergyman's sore throat" are experienced, even ordinary conversation being an effort.

At first the symptoms are merely annoying, but with each annual recurrence they become more and more severe, and life is rendered a perfect misery during the three brightest months of the year. The symptoms are associated with intense prostration, altogether out of proportion to the local irritation, and health is greatly impaired for many weeks after the peculiar symptoms have subsided.

Asthma is generally considered to be due to spasm of the involuntary muscular fibres of the bronchioles, but there is a growing tendency to give up this view and to regard the affection as a spasm of the vessels of the mucous membrane of the bronchi, due to an atonic condition of the vaso-motor nerves (Weber). In spasmodic asthma probably both of these factors are combined, for Lazarus has found as the result of irritation of the vagus in lower animals that the attack begins with broncho-spasm and stenosis, followed by the catarrhal symptoms usually observed. That asthma is frequently excited reflexly from peripheral irritation, *e.g.*, in the stomach from indigestible food, is well known, and need not be discussed here; but it has been shown that the abnormal conditions in the nasal passages which predispose to hay-fever, may reflexly excite attacks of spasmodic asthma. Thus in the worst cases of hay-fever attacks of spasmodic asthma come on early, and sometimes the patient is never free from asthma for weeks continuously. The asthmatic condition persists for some time after the hay-fever has disappeared, and the period of persistence tends every year to lengthen, till the unhappy sufferer becomes a confirmed asthmatic, and is never free from the attacks throughout the year.

Treatment.—As in all neuroses, we should bear in mind the importance of treating the general constitutional condition by promoting the health and strength of those who are subject to paroxysmal sneezing or asthma. For this purpose a general hygienic and tonic treatment should be adopted, and in hay-fever a tonic course is desirable for a few months before the usual time for the symptoms to come on.

If sneezing or asthma is due to abnormal conditions of the nasal passages, radical measures should be undertaken with a view to their removal.

When erectile swelling and vascular injection of the mucous membrane is the only abnormality, I have found that spraying the nasal passages with a solution of iodide of mercury (I invariably use "iodic-hydrarg.," B. W. and Co.), of strength 1 in 1000 to 1 in 250, has proved highly successful in curing many cases, myself among the number. A cocaine spray should be used beforehand, but as the cocaine is destroyed by the mercurial salt, it is necessary to relieve the pain which soon comes on by a hypodermic injection of morphine. The mucous membrane of the nose becomes much congested and swollen. In about three hours the pain and swelling subside, and are followed by a nasal catarrh lasting two or three days. In suitable cases, if this be efficiently done at the onset of the symptoms of hay-fever, the patient will remain free throughout the season, and there are very few people who have suffered from hay-fever who will not readily undergo this or any treatment which promises relief. This method has the advantage of leaving the sense of smell unimpaired, and involves no destruction of tissue. Sir Andrew Clarke advised a somewhat similar procedure, which gave relief in about 50 per cent. of his cases, viz., painting the nasal mucous membrane with a mixture of carbolic acid, quinine, and perchloride of mercury.

Hypertrophic rhinitis and sensitive spots should be cauterised with the galvano-cautery, while septal deviations, polypi, spurs, and other abnormalities should be removed.

Idiopathic paroxysmal sneezing dependent on functional affections of the nerve centre should be combated by nervine tonics, such as arsenic, strychnine, valerianate of zinc, and quinine.

Cocaine should never be locally applied for the relief of hay-fever; it only tends to aggravate the condition after its transient good effects have passed off. Inhalations of the vapor benzoini, or v. benzoini c. chloroformo, are very soothing, as also are creasote, camphor, pinol, terebene, etc. These latter, dissolved in liquid vaseline, may be sprayed into the nostrils.

During the paroxysm of hay asthma only palliative treatment can be resorted to, such as inhalation of nitre fumes, the administration of belladonna, stramonium, lobelia, nitro-glycerine, or iodide of potassium. For asthma a favourite remedy with me is a draught containing liquid extracts of quebracho, grindelia robusta, myrtus chekan, and yerba santa, twenty minims of each with or without a few grains of pot. iod., and given in hot water with a little brandy every hour till the attack is relieved.

Chapter XXV.

FOREIGN BODIES IN THE NOSE.

FOREIGN BODIES AND PARASITES.

Foreign Bodies.

VARIOUS foreign bodies, such as small pebbles, buttons, beans, pips, and pieces of slate pencil, are found at times in the nasal passages. Of course, if the patient is brought with the statement that some such body has been pushed into the nose, the diagnosis is easy enough, but very often months or years have elapsed, and meanwhile granulations may have sprung up from the irritated mucous membrane, concealing the foreign body, while the patient has completely forgotten having pushed it in, or it may have reached the nose in vomiting.

The usual symptoms are pain in the nose and frontal region, with *unilateral* muco-purulent discharge, which is often fœtid and sometimes streaked with blood from ulceration or from the exuberant soft granulations that not infrequently conceal the body from inspection.

The conditions with which a foreign body may be confused are malignant disease, osteoma, ozæna, lupus, tubercular disease, caries, and post-nasal adenoids.

Rhinoliths.

Any foreign body lying in the nasal passages for a lengthy period is liable to become encrusted with calcareous matter and form a rhinolith. In many rhinoliths no definite nucleus can be found, but in these it has usually formed round a nucleus of blood or mucus.

The symptoms and physical signs are practically the same as in other foreign bodies. They sometimes reach an enormous size, especially in tropical climates; thus Headley reports one which weighed 720 grains. They consist chiefly of carbonate and phosphate of lime, with about 30 per cent. of organic matter.

The treatment of foreign bodies consists in careful extraction by means of forceps, a bent probe, or other means which may suggest themselves in any particular case.

Parasites.

Various living creatures and fungi are found at times in the nasal passages, such as earwigs, ascarides, oidium albicans, aspergillus, and they may give rise to considerable irritation and some rhinitis; but there is one affection which is practically confined to tropical climates, in which the nasal passages are infested with maggots, the larvæ of the screw worm, Sarcophaga Georgina or Lucilia Hominivora, the affection being known by the name "Peenash."

"Peenash" usually attacks those who are affected with ozæna or catarrhal rhinitis. The symptoms are intense irritation of the nose, with agonizing pain in the nose and frontal region. There is profuse sanious muco-purulent discharge from the nose, with quantities of maggots escaping by the anterior and posterior nares, and sometimes from the ears. The subcutaneous tissues become inflamed and œdematous; generally symptoms of encephalitis supervene and the patient dies comatose.

The only treatment which offers good prospect of cure is the injection of pure chloroform into the nasal passages under the influence of an anæsthetic. If the maggots infest the nasal passages only, there is a prospect of cure; but when the accessory sinus or the subcutaneous tissues were infiltrated, Kimball found that even 50 per cent. of carbolic acid, and 1 in 500 solution of perchloride of mercury, useless, while pure oil of turpentine only killed a few of the maggots. He advises the use of carbolised oil injections to relieve pain after the injection of pure chloroform.

FORMULÆ.

INHALATIONS.
To be used with a Steam Inhaler.

1. ℞.—Compound Tincture of Benzoin 1 fl. dr., to a pint of Hot Water (104° F.) for each inhalation.
   Sedative.

2. ℞.—Compound Tincture of Benzoin 1 fl. dr., with Chloroform 2 to 3 ♏. Mix and add to a pint of Hot Water (104° F.).
   Sedative in painful laryngitis.

3. ℞.—Creasote 80 ♏, light Carbonate of Magnesia 30 grs., Distilled Water to 1 fl. oz. A teaspoonful in a pint of water for each inhalation.
   Stimulant and antiseptic. Useful in ozæna and chronic laryngitis.

4. ℞.—Oil of Scotch Pine, or Oil of Swiss Pine, 40 ♏, light Carb. of Magnes. 20 grs., Water to 1 fl. oz. Add to a pint of hot water, and inhale.
   Stimulant in chronic laryngitis.

PASTILS.

A good basis for Pastils is the following modified form of the Glyco-Gelatine Paste of the Throat Hospital, London:—

5. ℞.—Refined Gelatine 1 oz., Glycerine by weight $2\frac{1}{2}$ ozs., Gum Acacia 2 drs., Orange Flower Water, or Triple Rose Water, 2 fl. ozs. Isinglass makes a pastil which dissolves more slowly.

6. ℞.—Menthol $\frac{1}{4}$ to $\frac{1}{2}$ gr., Glyco-Gelatine Paste q. s.
   In irritable or painful laryngeal affections.

7. ℞.—Tartrate of Morphine $\frac{1}{20}$ to $\frac{1}{10}$ gr., Emetin (extractive) $\frac{1}{20}$ gr., Glyco-Gelatine Paste q. s.
   Sedative and expectorant.

8. ℞.—Hydrochlorate of Cocaine, $\frac{1}{20}$ to $\frac{1}{8}$ gr., Glyco-Gelatine Paste q. s.
   Sedative.

9. ℞.—Codeine $\frac{1}{8}$ to $\frac{1}{4}$ gr., Citric Acid $\frac{1}{2}$ gr., Elixir of Saccharin q. s., Glyco-Gelatine Paste q. s.
Very useful in allaying the irritable cough of phthisis.

10. R. Emetin (extractive) $\frac{1}{15}$ gr., Tincture of Tolu 3 ♏, Codeine $\frac{1}{6}$ gr., Oil of Cubebs 1 ♏, Elixir of Saccharin q. s.
Forty minims of the Elixir of Saccharin will sweeten 8 ozs. of pastils.

GARGLES.

11. ℞.—Tannic Acid 1 dr., Water to 4 fl. ozs.

12. ℞.—Borax 40 grs., Tincture of Myrrh 30 ♏, Glycerine 1 fl. dr. Water to 4 fl. ozs.

13. ℞.—Chlorate of Potash 30 grs., Alum Sulph. 30 grs., Sol. Morph. Hydrochl., 30 ♏, Dilute Hydrochloric Acid 60 ♏, Glycerine 4 fl. drs., Water to 4 fl. ozs.

14. ℞.—Tannic Acid 12 grs., Rectified Spirit 6 ♏, Camphor mixture to 1 fl. oz. (Th. H. Ph.).
Astringent.

15. R. Gallic Acid 120 grs., Tannic Acid 360 grs., Water 1 fl. oz.
Rub the acids, finely powdered, into a small quantity of water, then add water to one ounce.
For arresting hæmorrhage after excision of the tonsils.
The mixture may be slowly sipped if necessary (Th. H. Ph.).

Goddard's Astringent Gargle.

16. ℞.—Red Rose Leaves 2 drs., Dilute Sulphuric Acid $\frac{1}{2}$ fl. dr. Boiling Water 5 fl. ozs. Infuse and strain when cold then add Clarified Honey 1 oz., Tannic Acid 40 grs. Alum 2 drs., Rectified Spirit of Wine and Rose Water of each 6 fl. ozs.

17. ℞.—Hazeline 1 fl. dr., Chloride of Sodium 10 grs., Bora 5 grs., Distilled Water to 1 fl. oz.

MIXTURES.

18. ℞.—Hydrochlorate of Apomorphine $\frac{1}{20}$ gr., Syrup of Virginian Prune (B. P. C.) 30 ♏, Distilled Water to 1 fl. dr.
Every four hours.

19. ℞.—Ipecacuanha Wine 10 ♏, Tartrated Antimony $\frac{1}{50}$ gr., Cinnamon Water to 1 fl. dr.
    For laryngitis and capillary bronchitis, given at frequent intervals.

20. ℞.—Liquid Extract of Myrtus Chekan . . . . 20 ♏
    „ „ „ Yerba Santa . . . . . 20 ♏
    „ „ „ Grindelia Robusta . . . 20 ♏
    „ „ „ Quebracho . . . . . 1 fl. dr.
    Brandy . . . . 2 fl. dr.
    For spasmodic asthma, to be taken with half a tumblerful of hot water.

    Warburton Begbie's Mixture (Edin. Royal Inf. Ph.).

21. ℞.—Dilute Hydrocyanic Acid ½ fl. dr., Dilute Nitric Acid 3 fl. drs., Glycerine 1 fl. oz., Infusion of Quassia to 6 fl. ozs.
    A tablespoonful in a wineglassful of water three times daily. Sedative and tonic in phthisis.

INSUFFLATIONS.
Nasal.

22. ℞.—Subcarb. of Bismuth, ¼ gr., Acetate of Morphine $\frac{1}{32}$ gr, Powdered Starch ¼ gr.
    Make one powder for insufflation.

    Ferrier's Snuff.

23. ℞.—Hydrochlorate of Morphine 2 grs., Powd. Acacia Gum 2 drs., Subnitrate of Bismuth 6 drs.
    A small pinch to be insufflated at a time.

24. ℞.—Salicylic Acid (powd.) 10 grs., Tannic Acid (powd.) 60 grs., Subcarb. of Bismuth 60 grs.
    For nasal catarrh (Lefferts, New York).

25. ℞.—Iodoform in fine powder 3 grs., Tartrate of Morphine ¼ to ½ gr.
    One powder. Sedative and antiseptic in painful purulent conditions (Ed. Ryl. Inf. Ph.).

    Menthol Snuff.

26. ℞.—Menthol 1 part, Boric Acid (powd.) 2 parts, Chloride of Ammonium (powd.) 3 parts.
    Sedative and antiseptic.

27. ℞.—Borax (powder) 10 grs., Chloride of Sodium 20 grs., Chloride of Ammonium 10 grs., Camphor 1 gr.
Mildly stimulating in chronic rhinitis.

Laryngeal.

28. ℞.—Hydrochlorate of Morphine $\frac{1}{10}$ to $\frac{1}{4}$ gr., Oxychloride of Bismuth to 3 grs.
For painful affections of the larynx, especially tubercular laryngitis.

29. ℞.—Iodoform in fine powder 1 gr., Boric Acid 1 gr., Hydrochlorate of Morphine $\frac{1}{12}$ to $\frac{1}{6}$ gr.
For tubercular ulcers.

SYRUPS.

Compound Syrup of Camphor (Bristol Royal Infir. Phar.)

30. ℞.—Camphor 120 grs., Oil of Anise 2 fl. drs., Benzoic Acid, 180 grs., Glacial Acetic Acid 7 fl. ozs., Tincture of Opium 11 fl. ozs., Squills 5 ozs., Ipecacuanha 2 ozs., Purified Sugar 28 lbs., Burnt Sugar enough to colour, Distilled Water to 4 gallons.
One teaspoonful to be taken occasionally.

Syrup of Codeine.

31. ℞.—Codeine 40 grs., Dil. Phosph. Acid 1 fl. dr., Distilled Water 4 fl. dr., Rect. Spir. of Wine 4 fl. dr., Syrup to 20 fl. ozs. Dissolve and mix. One drachm = $\frac{1}{4}$ gr. (Edin. Ryl. Infirm. Ph.).
One teaspoonful to be taken when necessary.

Linctus Limonis Compositus.

32. ℞.—Hydrochlorate of Morphine $\frac{1}{2}$ gr., Dilute Hydrocyanic Acid 16 ♏, Spirits of Chloroform 12 ♏, Glycerine 2 fl. drs., Syrup of Lemons 2 fl. drs., Distilled Water to 1 fl. oz. One drachm equals $\frac{1}{16}$ gr. of Hydrochlorate of Morphine.
One teaspoonful to be taken in water when necessary.

PAINTS.

33. ℞.—Chromic Acid 45 grs., Distilled Water 1 fl. oz.
For application to ulcerated surfaces by means of a finely-pointed cotton-wool brush. To be followed quickly by an alkaline wash (Cent. Lond. Thr. Hosp.).

FORMULÆ. 273

34. ℞.—Nitrate of Silver 10 to 60 grs., Distilled Water 1 fl. oz.
35. ℞.—Sulphate of Copper 10 to 20 grs., Distilled Water, 1 fl. oz.
36. ℞.—Chloride of Zinc 10 to 30 grs., Hydrochloric Acid 1 ℳ, Distilled Water to 1 fl. oz.
37. ℞.—Perchloride of Iron (anhydrous) 30 to 120 grs., Distilled Water 1 fl. oz.

Mandl's Solution of Iodine (Schech).

38. ℞.—Iodine (pure) $6\frac{1}{4}$ to 20 grs., Iodide of Potassium 25 to 75 grs., Oil of Peppermint 3 ℳ, Glycerine to 1 fl. oz.
39. ℞.—Papain 4 grs., Lactic Acid 4 ℳ, Distilled Water to 1 fl. oz.

FOR USE WITH THE OIL ATOMISER.

40. ℞.—Eucalyptol 10 grs., Terebene (pure) 10 ℳ. Colourless Oil of Vaseline 1 fl. oz.
    Antiseptic.
41. ℞.—Cocaine 5 grs., Menthol 15 grs., Pure Terebene 10 ℳ., Colourless Oil of Vaseline 1 fl. oz.
    Sedative.

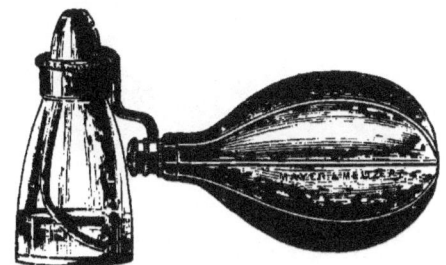

FIG. 119.
The Oil Atomiser.

42. ℞.—Eucalyptol 10 grs., Camphor 2 grs., Oil of Scotch Pine 8 ℳ, Colourless Oil of Vaseline 1 fl. oz.
    Stimulating in chronic rhinitis.
43. ℞.—Eucalyptol 10 grs., Aristol 20 grs., Menthol 20 grs., Colourless Oil of Vaseline 1 fl. oz.
    In fœtid rhinitis.

44. ℞.—Eucalyptol 15 grs., Menthol 10 grs., Camphor 2 grs., Morphine 4 grs., Cocaine 6 grs., Eucalyptus Oil 6 ♏., Colourless Oil of Vaseline 1 fl. oz.
   In acute rhinitis.

DOUCHES AND SPRAYS.

45. ℞.—Carbolic Acid 30 grs., Bicarbonate of Soda 15 grs., Borax 15 grs., Glycerine 45 ♏, Distilled Water to 1 fl. oz.
   To be used with an equal quantity of warm water.
   Spray up the nostrils with an atomiser or with the nasal douche. Mildly detergent, useful in fœtid discharge from the nose, and in hypertrophic rhinitis.

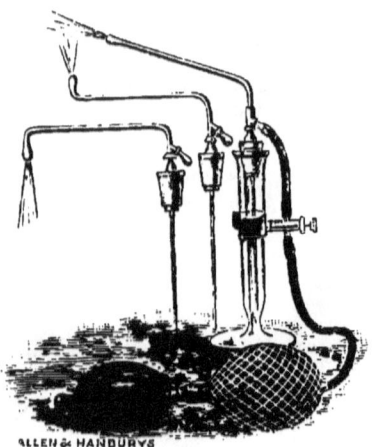

FIG. 120.

A convenient Spray for nose, naso-pharynx, and larynx. The rubber ball having been previously distended, the spray is started by compressing the button, and thus it can be worked with one hand.

46. ℞.—Listerine* 2 fl. drs., Borax 10 grs., Bicarbonate of Soda 10 grs., Water to 1 fl. oz.
   Useful for removing crusts of secretion from the nose.
   *A mixture of the essential antiseptic constituents of thyme, eucalyptus, baptisia, gaultheria, and mentha arvensis in combination.

47. ℞.—Sulpho-Carbolate of Zinc 2 grs., Distilled Water to 1 fl. oz.
   Antiseptic.

48. ℞.—Papain 15 grs., Lactic Acid 15 ♏, Distilled Water to 1 fl. oz.
For dissolving diphtheritic membrane.

49. ℞.—Aceto-Tartrate of Alum 1 oz., Distilled Water to 8 fl. ozs.
A tablespoonful to a pint of water to form a gargle, nasal douche or spray. Specially useful in ozæna (Edin. Ryl. Infirm. Pharm.).

50. ℞.—Salicylate of Soda 2½ drs., Borax 4 drs., Glycerine 4 fl. drs., Distilled Water to 8 fl. ozs.
To use 2 drs. in a pint of warm water.

51. ℞.—Hydrochlorate of Cocaine 30 grs., Salicylic Acid ½ gr., Distilled Water to 1 fl. oz.

GENERAL INDEX.

	PAGE
ABDUCTOR paralysis	169
Abscess of the antrum of Highmore	221
— — larynx	79
— — septum	218
— retro-pharyngeal	28
Accessory sinuses, diseases of the	221
Acute catarrhal pharyngitis..	21
— glanders	238
— inflammations of the larynx	73
— inflammatory œdema ..	79
— laryngeal obstruction ..	99
— laryngitis	73
— — in children	75
— phlegmon of the pharynx and larynx	23
— purulent rhinitis	191
— rhinitis	190
Adductor paralysis	170
Adenoids, post-nasal	205
Air, warmed by nasal passages	5
Anæmia of tubercular larynx	128
Anæsthesia, nasal	258
— of the larynx	156
— — pharynx	71
Anatomy and physiology ..	1
Angina Ludovici	26
Angioma of larynx ..	143, 147
— — nose	252
Ankylosis of the crico-arytenoid joint	173
Anosmia	256
Anterior rhinoscopy	183
Antitoxin treatment of diphtheria	97
Antral empyema	222
Antrum of Highmore, diseases of	221, 228
Aphonia	7
Aprosexia	244
Arytenoid cartilages, fixation of	173

	PAGE
Arytenoideus, paralysis of the	167
Arytenoiditis	74
Asthma	262
Atrophic pharyngitis.. ..	32
— rhinitis	198
Atrophy of the tonsils ..	52
BACILLUS of diphtheria, detection of the	94
Benign neoplasms of the fauces	64
— — — larynx	141
— — — nose	241
Bichromate of potash in chronic laryngitis ..	87
Blood-tumours of the nose 218, 220
Bulbar paralysis, larynx in 166, 177
CALCAREOUS concretions ..	64
— — in the nose	266
— — in the tonsil	52
Calomel fumigations in croup	93
Cancer, encephaloid	150
Carcinoma of the fauces ..	65
— in the nose	253
Catarrhal laryngitis	73
— pharyngitis	21
Cells, ethmoidal, diseases of the	229
Chicken pox, throat affections	116
Chondritis of the larynx ..	82
Chorditis	74
— tuberosa	88
Chorea, laryngeal	162
Chronic empyema of the antrum	221
— enlargement of the tonsil	48
— glanders	237
— hypertrophic rhinitis ..	194
— laryngitis	84
— pharyngitis..	30
— rhinitis	191
Congenital defects and deformities of the nose ..	189
Coryza caseosa	204

	PAGE
Coryza œdematosa	204
Cough, nasal	259
Crests of the septum	212
Crico-arytenoid joint, ankylosis of	173
Crico-thyroid paralysis	164
Croup, false	76
— membranous	90
Croupous rhinitis	203
Curettement of laryngeal tuberculosis	135
Cystic and other diseases of the antrum	228
Cystoma of larynx	143, 147
Cysts in the nose	253
DEFECTS and deformities, congenital, of the nose	189
Deflections of septum	212
Diagnosis of intra-laryngeal paralysis	178
— tables for differential	70, 255
Diagram of rhinoscopic image	187
Differential diagnosis, tables for	70, 255
Difficulties in laryngoscopy	18
Diphtheria, laryngeal	96
— nasal	96
Diphtheritic exudation in typhoid fever	107
— — — measles	114
— — — scarlatina	113
— — — smallpox	115
Diseases of the accessory sinuses	221
— — — antrum	221, 228
— — — ethmoidal cells	229
— — — frontal sinus	229
— — — larynx	73
— — — nose	190
— — — pharynx	21
— — — septum	212
— — — sphenoidal sinuses	231
— — — tonsils	44
Disseminated sclerosis, larynx in	177
Douches and sprays	274
Dysphonia spastica	161
EAR, affections of, from post nasal adenoids	206
Ecchondroma of larynx	143
Ecchondroses in the nose	252

	PAGE
Elongated uvula	39
Empyema, chronic, of the antrum	221
Enteric fever, throat affections	107
Enucleation of the tonsil	50
Epiglottis, epithelioma of the	149
Epiglottitis	74
Epistaxis	218
Epithelioma of larynx	148
— — pharynx	64
Erysipelas of the pharynx and larynx	23, 26
Ethmoidal cells, diseases of the	229
Examination of the pharynx and larynx	10
Exostoses in the nose	253
Expiratory glottic spasm	161
FALSE croup	76
Fauces, sarcoma of the	65
Fever, hay	260
Fevers, throat affections of infectious	107
Fibrinous rhinitis	203
Fibroid degeneration of laryngeal cartilages	82
Fibroma in the nose	252
Fibrous polypus of larynx	142
Fish bones in larynx	181
Fixation of cords	124
Forceps, laryngeal, Cusco's	181
— — Dundas Grant's	146
— — Mackenzie's	146
— — Schrötter's	146
— — Wolfenden's	181
— nasal polypus	247
Foreign bodies in the larynx	181
— — — nose	266
Formulæ	269
Frontal sinus, diseases of the	229
Fumigations, calomel, in croup	93
Functions of the nose	2
GALVANO-CAUTERY	36
Galvano-puncture of the tonsil	50
Gargles	270
General semeiology	6
Glanders	237
Gouty affections of the throat	117
Granular laryngitis	86

HÆMORRHAGE from the nose 218
— — tonsil after tonsillotomy 51
Hay asthma 265
— fever 260
Hyperæsthesia, nasal .. 258
— of the larynx 156
Hypertrophy of the pharyngeal tonsil 205
— — inferior turbinated bodies 195

INFLUENZA, throat affections 116
Inhalations 269
Inspiratory glottic spasm .. 159
Insufflations 271
Intra-laryngeal injections 87, 137
Intubation of the larynx .. 101

LABIO-GLOSSO-LARYNGEAL paralysis, the larynx in 177
Laryngeal catarrh, acute .. 73
— — chronic.. 84
— — syphilitic 122
— — tubercular 127
— chorea 162
— complications in infectious fevers 107
— crises 176
— diphtheria 96
— manifestations of chronic diseases of the central nervous system .. 175
— obstruction, acute .. 99
— ulcers, catarrhal 86
— — lupous 60
— — syphilitic 122
— — tubercular 129
— — typhoid.. 108
— vertigo 163
Laryngismus stridulus .. 159
Laryngitis, acute 73
— chronic, intralaryngeal injections for .. 84, 87
— granular 86
— hæmorrhagica .. 74, 86
— herpetica 74
— hypoglottica 74
— sicca 74
— spasmodic 76
— stridulosa 76
— tubercular 127
Laryngoplegia 171
Laryngorrhœa 85

Laryngoscopy.. .. 10, 14
— difficulties in 18
Laryngo-tracheal ozæna .. 200
Larynx, acute inflammations of the 73
— anæsthesia of the 156
— and pharynx, acute phlegmon of the 23
— — — cancer of the .. 148
— — — congenital web of the 20
— — — erysipelas of the .. 23
— — — examination of .. 14
— — — lupus of the .. 54
— — — neoplasms of the .. 64
— chondritis of the 82
— foreign bodies in the .. 181
— hyperæsthesia of the .. 156
— in influenza 116
— in typhoid fever 107
— intubation of the 101
— leprosy of the 137
— malignant neoplasms of .. 148
— motor neuroses of .. 157
— neoplasms of the 141
— neuroses of the 156
— œdema of the 79
— papilloma of 142
— perichondritis of the .. 82
— sarcoma of the 150
— syphilis of the 121
— trans-illumination of the.. 20
Leprosy of the throat and nose 137
— tuberculated 138
Lipoma and adenoma of larynx 144
Locomotor ataxia 175
Lupous ulceration 68
— ulcers, diagnosis of .. 67
Lupus of the nose 36
— — pharynx and larynx .. 59
Luxation of the crico-arytenoid joint 175

MALIGNANT neoplasms of pharynx and fauces 64
— — in the nose 253
— — of larynx 148
— — — — diagnosis of .. 151
— ulceration 69
— ulcers, diagnosis of .. 67
Measles, throat affections in 114

GENERAL INDEX.

	PAGE
Membranous croup	90
— rhinitis	203
Mixtures	270
Mogiphonia	161
Motor neuroses of pharynx	71
— — of larynx	157
Mouth-breathing	3
Mucous polypus	241
Muscle, paralysis of the crico-thyroid	166
— supplied by the superior and recurrent nerves	167
Muscles supplied by the superior laryngeal nerve	164
— paralysis of laryngeal	163
— supplied by the recurrent laryngeal nerves	167
Mycosis, pharyngeal	62
Myxoma of larynx	143
NASAL anæsthesia	258
— cough	259
— diphtheria	96
— douche	202
— hyperæsthesia	258
— neuroses	256
Naso-pharyngitis	211
Naso-pharynx, neoplasms of the	241
Neoplasms	123
— benign of pharynx	64
— malignant of pharynx	64
— of the larynx, benign	141
— — — malignant	148
— — nose	241, 253
— — pharynx and larynx	64
Neuroses, nasal	256
— of the larynx	156
— — pharynx	71
Non-inflammatory œdema of larynx	81
Nose, leprosy of the	137
— cysts in the	253
— ecchondroses of the	253
— exostoses of the	253
— fibroma of the	252
— foreign bodies in the	266
— functions of the	2
— hæmorrhage from the	218
— lupus of the	236
— neoplasms of the	241
— osteoma of the	253
— papilloma of the	252

	PAGE
Nose, sarcoma of the	253
— syphilis of the	232
— tuberculosis of the	234
OBSTRUCTION to respiration	8
O'Dwyer's operation for acute laryngeal stenosis	101
— — — chronic syphilitic stenosis	126
Odynphagia	8, 9
Œdema of the larynx	79
— — — — angio-neurotic	81
— — — — in influenza	116
— — — — in typhoid fever	111
Olfactory neuroses	256
Operations, radical, of larynx	153
Ossification of laryngeal cartilages	82
Osteoma in the nose	253
Ozæna	198
— laryngo-tracheal	200
PACHYDERMIA diffusa	85
— laryngis	88
Papilloma of larynx	142
— — the nose	252
Paralysis after section of vagus nerve	179
— diagnosis of intra-laryngeal	178
— labio-glosso-laryngeal	177
— of laryngeal muscles	163
— in laryngeal tuberculosis	129
— of the abductors of the vocal cords	168
— — adductors of the vocal cords	170
— — arytenoideus	167
— — crico-thyroid muscle	166
— — sphincters of the glottis	164
— thyro-arytenoidei interni	165
Parasites in the nose	266, 267
Paresis of laryngeal muscles	163
Parosmia	257
Paroxysmal sneezing	260
Pastils	269
Perforation of the septum	212
Perichondritis of the laryngeal cartilages	82, 123
— — — — in enteric fever	111
Pharyngeal tonsil, hypertrophy of the	205

	PAGE
Pharyngitis, acute catarrhal	21
— atrophic	32
— chronic	30
— — granular	34
— gouty	117
— sicca	35
Pharyngo-mycosis leptothricalis	62
Pharyngoscopy	11
Pharynx, syphilis of ..	54
— tuberculosis of	57
— and larynx, acute phlegmon of the ..	23
— — — erysipelas of the ..	23
— — — examination of	10
— — — lupus of the	59
— — — neoplasms of the ..	64
— neuroses of the	71
— spasm of the	72
Phonic spasm ..	161
Physiognomy in syphilitic disease	9
— — post-nasal growths	206
Polypi, snaring	248
Polypus, fibrous, of larynx ..	142
— mucous of nose	241
Posterior rhinoscopy..	186
Post-nasal adenoids ..	205
Primary epithelioma ..	64
RADICAL operations of larynx	153
Respiration, obstruction to ..	8
Retro-pharyngeal abscess ..	28
Rheumatic affections of the throat ..	119
Rhinitis, acute	190
— acute purulent	191
— atrophic	198
— caseosa	204
— chronic	191
— — hypertrophic	194
— — purulent	192
— croupous	203
— fibrinous	203
— membranous	203
— oedematosa	204
— sicca	192
— vaso-motor	259
Rhinoliths	266
Rhinoscleroma	239
Rhinoscopic image, diagram of	188
Rhinoscopy	183

	PAGE
SACCULUS laryngis	17
— — prolapse of	144
Sarcoma of the nose	253
— — larynx ..	150
— — the fauces	65
Scarlatina, throat affections in	112
— anginosa	112
Sclerosis, disseminated, larynx in	177
Semeiology, general ..	6
Sensory neuroses of the pharynx	71
— — — larynx	156
Septal deflections	212
Septum, diseases of the	212
— perforation of the..	212
Silent croup ..	91
Sinus, frontal, diseases of the	229
Sinuses, accessory, diseases of the ..	221
— sphenoidal, diseases of the	231
Small pox, throat affection in	115
Snaring polypi	248
Sneezing, paroxysmal	260
Snuffs ..	271
Spasm, expiratory glottic	161
— inspiratory glottic	159
— — in adults	160
— of the pharynx	72
— phonic	161
Spasmodic laryngitis	76
Sphenoidal sinuses, diseases of the ..	231
Sphincters of the glottis, paralysis of the	164
Sprays and douches ..	274
Spurs of the septum ..	212
Stammering of vocal cords ..	161
Stenosis of larynx, chronic syphilitic	123
Superficial syphilitic ulcer, the	67
— ulceration ..	55
Syphilis of the pharynx	54
— — larynx	121
— — nose..	232
Syphilitic catarrh	122
— laryngeal stenosis..	123
— neoplasms ..	123
— perichondritis of larynx	82, 123
— ulcer, the deep	68
— ulcers, diagnosis of	67

	PAGE
TABLES for differential diagnosis	70, 255
Tensors, internal, of the vocal cords	165
Throat affections of infectious fevers, etc.	107
Thuja in laryngeal papilloma	147
Thyro-arytenoid paralysis	165
Tonsil, abscess of the	45, 48
— chronic enlargement of the	48
— concretions in the	52
— enucleation of the	50
— galvano-puncture of the	50
— hypertrophy of the pharyngeal	205
— the lingual	53
Tonsillitis	44
Tonsils, atrophy of the	52
— diseases of the	44
— hæmorrhage of the	51
— malignant disease of the	64
Tracheal probang for false membrane in larynx	100
Transillumination in diagnosis of antral disease	224
— of larynx	20
Tubercular laryngitis	127
— ulcers, diagnosis of	67, 70
Tuberculated leprosy	138
Tuberculosis of the pharynx and fauces	57
— — — nose	234

	PAGE
ULCERATION, lupous	68
— malignant	69
— tertiary syphilitic	55
— tubercular of larynx	129
— — — pharynx	57
— — — nose	234
Ulcers, differential diagnosis of	67
— tubercular, of the nose	235
Unilateral paralysis of left and right cord	173
— — — adductor of vocal cords	171
Uvula, elongated	39
VAGUS nerve, paralysis after section of	179
Valleculæ	17
Varix and adenoid hypertrophy of base of tongue	43
— turbinal	194
Vaso-motor rhinitis	259
Ventricle of Morgagni	17
— — — prolapse of	144
Vertigo, laryngeal	163
Vocal cords, external tensor of	165
— — internal tensors of	166
— — paralysis of the abductors of the	168
— — — — adductors of	170
— — ulcers of, catarrhal	86
— — — syphilitic	122
— — — tubercular	130

A SYNOPSIS OF
THE PRACTICE
OF MEDICINE

FOR PRACTITIONERS AND STUDENTS.

By WILLIAM BLAIR STEWART, A.M., M.D.,

Lecturer on Therapeutics; late Instructor on Practice of Medicine in the Medico-Chirurgical College of Philadelphia; Demonstrator in the Philadelphia School of Anatomy, etc.

This work has been undertaken after several years of experience by [the] Author as Instructor on the subject of the Practice of Medicine, his [purp]ose being to prepare and present to the profession, a brief synopsis of [the] subject, not with the view of replacing the expensive and elaborate [publ]ications, but to give to the busy practitioner and student, at a small [cost,] concise and accurate descriptions which will suggest outlines and [prac]tical thoughts upon etiology, symptomology, pathology, diagnosis, [prog]nosis and treatment.

The author has used every endeavor to obtain the best material from [ever]y reliable source. All of the prominent authorities in the recently [issue]d Text-books and Systems, also the current Medical Literature, have [been] laid under contribution, and the most approved methods of treatment [have] been given prominence. Many drugs and methods have not been [cons]idered at length, not on account of their inutility, but from the fact [that] better forms of treatment have taken their place.

The Publisher calls the attention of the Medical Profession to this [wor]k; confident in the belief that both the practitioner and student, desir[ing] a Modern Synopsis of Practice, will not only find it all that the Author [and] Publisher claim for it, but far superior to any work of its kind.

[Jou]rnal of American Medical Association says:—" This book, as its title indi[cates], summarizes in a careful and useful [mann]er the existing practice. The treat[ment] recommended is sound and in accor[danc]e with the latest approved teachings."

[Th]e Medical Bulletin says:—" Dr. [Stew]art has certainly succeeded in com[press]ing into a volume of moderate size a [mass] of reliable information representing [his o]wn experience and the writings of the [best] authors upon the subject."

The Virginia Medical Monthly says:— "This handsome volume is what its title claims for it. It is about as concise a compilation from good authorities as could be well made, and is, therefore, a valuable book with which to review what has been learned from the authoritative works on practice. It is also valuable to the practitioner, who may be compelled to look in hast for leading facts to guide him in diagnosis, therapeutics, etc."

One Large Octavo Volume, 434 pages, Cloth, **$2.75.**

MANUAL OF
CLINICAL DIAGNOSIS.

BY

ALBERT ABRAMS, M.D.,

Professor of Pathology, Cooper Medical College, San Francisco, California; Pathologist to the City and County Hospital, San Francisco, California; Author of "A Synopsis of Morbid Renal Secretion," etc.; President of the San Francisco Medico-Chirurgical Society (1893-1894), etc., etc.

The study of the changes wrought by disease constitutes the scienc and art of Diagnosis, without which rational treatment is out of th question.

Hence it is the duty of the physician to thoroughly investigate th phenomena of the disease he is called to treat; in this way he arrives at knowledge of the causes, to which his skill in treatment should be directed and it is only by a perfect understanding of the changes which diseas produces in the body that the proper treatment can be adduced.

In this book the author has outlined and carefully described in th most concise and accurate manner possible, the signs and symptoms c diseases, and their importance to the diagnostician. His work on thi line shows him to be the master of his subject.

The book is well arranged, plain and easy of comprehension—much c the matter being arranged in tabular form—and is up to date in it methods.

University Medical Magazine says:— "The work contains a vast amount of valuable material, which has been carefully classified and condensed within a small compass."

American Therapist says:—"A very complete little book on physical diagnosis. The articles are concise and well written."

Pacific Medical Journal says:—"The old practitioner from its careful perusal will learn of the vast improvements in the methods of physical diagnosis since his college days."

The Journal of the American Medical Association says:—"The writer has produced a compact, useful manual."

Southern California Practitioner says —"The writer has evidently put muc time on the preparation of this book, an he has well classified a large amount c material into a compact whole."

Memphis Medical Monthly says:— "The practitioner and student will fin the arrangement of this book to be e) tremely satisfactory. We ha.e only word of commendation for this excellent littl volume."

Occidental Medical Times says:—" Th book must be studied to be fully apprec ated, and we do not know of a better inves ment for the practitioner as well as the stu dent for whom it was primarily intended.

In One Octavo Volume, Illustrated. Price, $2 75.

Nose and Throat.
Diseases of the Upper Respiratory Tract.
By P. WATSON WILLIAMS, M.D., M.R.C.S., (Lond.)

Physician in charge of Throat Department, Bristol Royal Infirmary; Honorary Physician to the Institute for the Deaf and Dumb.

CONTENTS—CHAPTERS I–XXV.

I. INTRODUCTION.—Anatomy and Physiology; General Semeiology.

II. EXAMINATION OF THE PHARYNX AND LARYNX.—Pharyngoscopy; Laryngoscopy.

III. ACUTE PHARYNGITIS.—Acute Catarrhal Pharyngitis; Acute Phlegmon and Erysipelas of the Pharynx and Larynx; Retro-pharyngeal Abscess.

IV. CHRONIC PHARYNGITIS.—Chronic Catarrhal, hypertrophic and atrophic pharyngitis; elongated uvula; varix and adenoid hypertrophy of the tongue.

V. DISEASES OF THE TONSILS.—Acute tonsilitis; Cronic Hypertrophy of the Tonsils; Atrophic Tonsilitis.

VI. CHRONIC INFECTIVE DISEASES.—Syphilis, Tuberculosis; Lupus of the Pharynx and Larynx, Mycosis.

VII. NEOPLASM OF THE PHARYNX AND FAUCES.—Benign Neoplasm; Malignant Neoplasm; Differential Diagnosis of Syphilitic, Tubercular, Lupous, and Malignant Ulcers.

VIII. NEUROSES OF THE PHARYNX.—Sensory Neuroses; Motor Neuroses.

IX. ACUTE INFLAMMATIONS OF THE LARYNV.—Acute Catarrhal Laryngitis; Spasmodic Laryngitis; Oedema Perichondritis of the Laryngeal Cartileges.

X. CHRONIC LARYNGITIS.—Chronic Laryngitis and Pachydermia Laryngitis; Chroditis Tuberosa.

XI. MEMBRANOUS CROUP AND DIPHTHERIA.—Membraneous Croup; Diphtheria; Intubation.

XII. THROAT AFFECTIONS OF INFECTIOUS FEVERS, GOUT AND RHEUMATISM.—Enteric Fever, Scarlatina, Measles, Smallpox, Chicken-pox, Influenza, Gout and Rheumatism.

XIII. CHRONIC INFECTIVE DISEASES.—Syphilis; Tuberculosis, Leprosy.

XIV. NEOPLASM OF THE LARYNX.—Benign Neoplasm; Malignant Neoplasm.

XV. NEUROSES OF THE LARYNX.—Sensory Neuroses; Motor Neuroses.

XVI. FOREIGN BODIES IN THE LARYNX.

XVII. RHINOSCOPY.—Anterior Rhinoscopy; Posterior Rhinoscophy.

XVIII. RHINITIS.—Acute Rhinitis, Chronic Rhinitis, Hypertrophic Rhinitis, Atrophic Rhinitis, Rhinitis Aedematosa and Caseosa.

XIX. HYPERTROPHY OF THE PHARYNGEAL TONSIL.—Post-Nasal Adenoids.

XX. DISEASES OF THE SEPTUM.—Perforation, Deflections and Spurs, etc. Epistaxis, Abscess.

XXI. DISEASES OF THE ACCESSORY SINUSES.—The Antrum of Highmore; The Ethmoidal Cells; The Frontal Sinus; The Sphenoidal Sinus.

XXII. CHRONIC INFECTIVE DISEASES.—Syphilis; Tuberculosis; Lupus; Glanders; Rhinoscleroma.

XXIII. NEOPLASMS OF THE NOSE AND NASOPHARYNX.—Mucous Polypus, and Benign Neoplasms; Malignant Neoplasms.

XXIV. NASAL NEUROSES.—Olfactory Neuros *;* Paræsthesiæ; Nasal Cough; Vaso-Motor Rhinitis; Hay Fever.

XXV. FOREIGN BODIES IN THE NOSE.

APPENDIX. Formulæ for Inhalations; Pastils; Gargles; Mixtures; Insufflations; Syrups; Paints; For use with the Oil Atomizer; Douches and Sprays.

One Octavo Vol.: Illustrations in Lithographic Colors, and in Black, Uniform with Medical Classics Series, Cloth, $2.75.

E. B. TREAT, Publisher, 5 Cooper Union, New York.

SURGICAL HANDICRAFT.

—A MANUAL—
OF
Surgical Manipulations, Minor Surgery, and other matters connected with the work of House Surgeons and Surgical Dressing.

By WALTER PYE, F.R.C.S.,

Surgeon to St. Mary's Hospital and the Victoria Hospital for Sick Children; late Examiner in Surgery at the University of Glasgow.

UPWARDS OF 300 ILLUSTRATIONS ON WOOD.

FIRST AMERICAN EDITION
FROM THE THIRD LONDON EDITION,

Revised and Edited by T. H. R. CROWLE, F R.C.S.,

Surgical Register to St. Mary's Hospital, and Surgical Tutor and Joint Lecturer on Practical Surgery in the Medical School.

"The subjects cover the whole range of surgery. The book is an exceedingly good guide for those for whom it seems to be intended, and it furnishes not only them but the general medical practitioner with an excellent book of reference for any emergency which may arise. The matter is well chosen, the style is clear and pleasant, the illustrations are many of them new, and the methods of treatment described are fully up to date. It is different in its scope from any book on minor surgery with which we are acquainted, and it compares favourably with them all.—*The New York Medical Journal.*

Edinburgh Medical Journal says:—"Such a rapid sale of the first edition is in itself a strong testimony to the favour with which the work has been received by the profession. The book can be recommended as a really useful one, containing much practical information, described in a clear and interesting way. Recent authorities have been consulted, so as to bring the matter well up to date. The illustrations are excellent, and have all the vividness of good life portraits. The presence of special chapters on subjects not usually treated of in works of this kind is a distinct advantage."

British Medical Journal says:—"We had occasion to notice favourably the first edition of this manual. The work is a very practical and useful one, and the author is to be congratulated upon its success. The student, too, will not be sorry to find the present edition less costly, as well as more full of information, and also more portable. The book is certainly worthy of its author's reputation, and can be safely recommended."

London Lancet says:—"Mr. Pye has reduced the size of his volume considerably, and has made some alterations in the text by excission and addition; he has thus added to it value, and also put it within the reach of a much wider circle of readers."

London Medical Record says:—"A new edition of this practical work, in so short a time after the first, is strong evidence of its appreciation by the public, and is a source of congratulation to the author, who has materially added to the usefulness of the book by slightly reducing its bulk, and yet, by using more condensed type, has been enabled to put more material into a smaller compass. Mr. Pye has made judicious use of recent additions to surgical literature, and the *Manual of Surgical Handicraft* is likely to prove a useful and popular addition to the student's library."

In one 8vo, vol. 600 pages. Illustrated. Cloth, $3.50, net; Sheep. $4, net.

E. B. TREAT, Publisher, 5 Cooper Union, New York.

www.ingramcontent.com/pod-product-compliance
Lightning Source LLC
Chambersburg PA
CBHW022048230426
43672CB00008B/1104